Science and Sensibility

The publisher gratefully acknowledges the generous support of the August and Susan Frugé Endowment Fund in California Natural History of the University of California Press Foundation.

Science and Sensibility

Negotiating an Ecology of Place

Michael Vincent McGinnis

UNIVERSITY OF CALIFORNIA PRESS

University of California Press, one of the most
distinguished university presses in the United States,
enriches lives around the world by advancing scholarship
in the humanities, social sciences, and natural sciences. Its
activities are supported by the UC Press Foundation and
by philanthropic contributions from individuals and
institutions. For more information, visit www.ucpress.edu.

University of California Press
Oakland, California

Library of Congress Cataloging-in-Publication Data

McGinnis, Michael Vincent, 1962- author.
 Science and sensibility : negotiating an ecology of place /
[Michael Vincent McGinnis].
    pages cm
 Includes bibliographical references and index.
  ISBN 978-0-520-28519-4 (cloth : alk. paper) — ISBN
0-520-28519-0 (cloth : alk. paper) — ISBN 978-0-520-
28520-0 (pbk. : alk. paper) — ISBN 0-520-28520-4
(pbk. : alk. paper) — ISBN 978-0-520-96075-6 (ebook)
— ISBN 0-520-96075-0 (ebook)
   1. Human ecology. 2. Human ecology--Case studies.
3. Environmental protection--Social aspects. I. Title.
 GF50.M38 2016
 304.2—dc23   2015032077

Manufactured in the United States of America

25   24   23   22   21   20   19   18   17   16
10   9   8   7   6   5   4   3   2   1

In keeping with a commitment to support
environmentally responsible and sustainable printing
practices, UC Press has printed this book on Natures
Natural, a fiber that contains 30% post-consumer waste
and meets the minimum requirements of ANSI/NISO
Z39.48–1992 (R 1997) (Permanence of Paper).

To my Mother, the sea, and my Father, the mountains

Contents

Preface

If you stay in a place long enough, you can begin to learn to listen to the landscape and the seascape that you inhabit. In June 2012 a group of scientists, writers, activists, and I gathered in Honolulu harbor, Oahu, and boarded an old sailing vessel. The vessel was a replica of a Polynesian *vaka*. Before we set sail we gathered on deck of the vaka in a circle. We held hands, and the navigator began to sing a song in Fijian. It was a song that celebrated a life with the sea and that invoked the power or *mana* for a safe voyage. The navigator of the vaka asked us to join in a chorus of song of appreciation for the sea. We all began to sing. We spoke different languages, so this was a difficult task. At first there was very little harmony, but in time our voices converged as if to capture a single great breath of the sea. *Aloha*. Aloha is the Pacific's song. A song of a common breath emerged and a chorus emerged which was followed by a great laughter. The laughter became the song, and our smiles marked a shared appreciation for our love of the sea. So we set sail with respect.

The chapters in this book emphasize the need for a deeper appreciation of our place in the world. The chapters are based on my personal journey and my experience with diverse coastal and maritime places and peoples across the Pacific Ocean. I have learned over time that there is one ocean that connects diverse peoples across the Pacific Ocean. We need to cultivate ecologically grounded values that can contribute to a science and sensibility of place. To re-inhabit a place and community can represent a first step in beginning to respond to the ecological threats and impacts we face in society.

While science is a key part of forging a more adaptive and resilient society, the cultivation of a renewed sense of place and community is essential to respond to the complex socio-ecological problems we face. This is my bioregional message. The book notes that modern science is one way of knowing, but there are other ways, other epistemologies, and other values that contribute to a practice of place-based living. There are other forms of knowledge that are as important, including local knowledge and traditional ecological knowledge systems.

The chapters in this book are the product of my research and writing, begun in 1992, on the challenges of protecting the health and integrity of watersheds, river basins, and marine ecosystems. I edited a book entitled *Bioregionalism* that was published by Routledge in 1999. This work led to studies of watershed activists and organizations, funded by three grants from the Ethics and Values Studies Program of the National Science Foundation (1993–1994, 1995–1997, 1998–2000). I thank Rachelle Hollander, who was at that time the director of the program, for her support. Later, in 2008, as one of the first Fulbright scholars to the Republic of Montenegro, I gave presentations on coastal and marine management across the Mediterranean Basin. In 2010 I began a faculty position at Victoria University of Wellington (New Zealand), and completed a comprehensive report, funded by the country's ministries, that included recommendations on how to strengthen New Zealand's ocean governance framework. My work in New Zealand culminated in a report on ocean governance in New Zealand for their environmental ministries in 2013. The chapters in this book draw from these experiences, and a number of publications based on research funded by the U.S. Department of the Interior, and my work for the National Marine Sanctuaries Program as well.

This book is the product of over fifteen years of discussions with scholars, professionals, and students. I would like to thank John Woolley, my good friend and a collaborator on several projects. I appreciate the assistance of James Binaski for his creative production of the artwork and graphics for this book. I also am grateful to ideas and intellectual feedback from a number of close colleagues and friends, including Ilya Ahmadizadeh, Linda Fernandez, David Schlosberg, Jonathan Boston, Jason Scorse, Bron Taylor, Freeman House, David Simpson, and the undergraduate and graduate students that I have worked with as an academic during the past twenty years. Richard Borden, Richard Howarth, and two anonymous reviewers also provided valuable suggestions and input on previous drafts of this book. An earlier version of chapter 2 was developed for an unpublished essay supported by the Santa Barbara Foundation. I appreciate the friendship of Sharyn Main and her thoughtful comments and input that went into this chapter. I would like to thank Charles Dawson and Teresa Shewry for their useful comments on a draft of chapter 4. Chapter 7 was based on funding from the Emerging Issues Program (2010–2012), overseen by the Institute for Policy Studies at Victoria University of Wellington (New Zealand). This chapter is also based on the feedback received during presentations on this subject that were given by the author in the United States and New Zealand in 2010–2013. In particular, I thank the participants in the 1st International Marine Conservation Think Tank, held in Auckland in December 2011. I also want to acknowledge the helpful information and materials provided by Sean Hastings of the National Marine Sanctuaries Program that contributed to the writing of chapter 8. I would like to thank Kelsey Richardson, who was my graduate student research assistant, and who provided an incredible level of intellectual support that contributed to the writing of chapter 9. I appreciate the faculty and students at the University of California, Santa Barbara, who invited me to participate in their seminar on Blue Justice, which was held on Santa Cruz Island in the spring of 2014. Special thanks to Kennedy Warren, who inspired many of the ideas expressed in chapter 9. I appreciate my editors Blake Edgar and Merrik Bush-Pirkle at University of California Press for their encouragement to write the book, and Roy Sablosky for his careful copyediting.

CHAPTER I

Negotiating Ecology in an Age of Climate Change

Ecology is nothing but this: the evaluation of place.
—Eduardo Viveiros de Castro (2013, 33).

The machine is stuttering and the engineers are in panic. They are wondering if perhaps they do not understand it as well as they imagined. They are wondering whether they are controlling it at all or whether, perhaps, it is controlling them.
—*Uncivilisation: The Dark Mountain Manifesto*

Imagine that as you watch the day go by you are sitting on your front porch whittling on a stick with a knife. With each whittle the stick is reduced. The wood shavings are at your feet. There is barely a stick left in your hand, and the shavings are swept up and used as kindling to light a fire that night. The whittling stick is a metaphor for how nature is whittled away by our continued use and abuse of the planet's ecosystems. The impacts of this whittling away of nature are often difficult to perceive and realize. For instance, the impacts of global climate change and other human impacts on nature and society are, at first glance, hard to recognize. Springtime may arrive late, or the sounds of spring birds may be missing from a forest. In his Nobel Prize–winning book *Raga*, Le Clézio describes the ocean as an "invisible continent" with its life flourishing unperceived or unnoticed. Akin to the sea life under the surface of the blue horizon, ecosystems are a mere shadow of their referential state. The natural world seems to be receding like a mirage in the Arizona desert. We are whittling away at the great circle of animals, plants, and insects that were once part of our communities.

"Precisely at the moment when we have overcome the earth and become unearthly in our modes of dwelling" (Harrison 1996, 428), we need to restore our kinship with the animate world and the places we inhabit. We are disabled creatures dislocated in a wounded landscape. Species loneliness in a wounded landscape moves us to want to restore our relationship with place and others, or, to put it another way, modern humanity yearns to re-establish and restore an ecology of shared identity. Rather than understanding the world through a relationship with earthly entities, our culture emphasizes the human ability to experience nature as a quality (or quantity) that springs from scientific, technological, bureaucratic, and economic understanding. Human beings remain isolated actors in an earthly cage; the world is technologically divided, scientifically categorized and manipulated, and perceived as absent of spiritual and intrinsic worth. Yet, the natural world is something more than the image depicted on the television or computer screen. Nature is more than an environment to behold, control, and manage.

How can we adapt to the social and ecological changes brought on by climate change if we hardly notice the changes? This chapter explores the wicked characteristics of global climate change, and the challenge of adaptation. Cultural adaptation requires a deep connection to place and one's region; it requires an understanding of the uniqueness of particular places that human beings are dependent on.

One consequence of industrialization is that we have created a "secondary nature"—a nature transformed by our use of it, by our technologies and machines, and by our behavior. Indeed, the science of ecology cannot tell us what is *natural* about ecosystems. The meanings of *nature* and *place* have significantly changed over time. Even the term *landscape* has lost its meaning. The root meaning of *landscape* is a forest stripped of trees, a hilltop cleared of native brush, a place where the natural terrain was removed and settled. *The natural landscape* is an oxymoron (Park 2006, 9). We inhabit landscapes transformed; so there is less nature or nativity to draw from, less to sense and perceive. This is particularly the case in marine systems, as marine scientist Jeremy Jackson (2001, 5411) explains:

> The persistent myth of the oceans as wilderness blinded ecologists to the massive loss of marine ecological diversity caused by overfishing and human inputs from the land over the past centuries. Until the 1980s, coral reefs, kelp forests, and other coastal habitats were discussed in scientific journals and textbooks as "natural" or "pristine" communities with little or no refer-

ence to the pervasive absence of large vertebrates or the widespread effects of pollution. This is because our concept of what is natural today is based on personal experience at the expense of historical perspective. Thus, "natural" means the way things were when we first saw them or exploited them, and "unnatural" means all subsequent change.

For instance, polar bears of the Arctic face a secondary nature in the ice ecology, and their abundance and distribution will likely diminish as a consequence (Post et al. 2013). As the Arctic ice retreats, polar bears are changing their behavior. In some areas, white-beaked dolphins are moving to the northern pole where they have not been seen before. Polar bears have shifted from those species that are ice dependent, such as ringed seals, to these white-beaked dolphins (Aars et al. 2015). Polar bears are also moving farther north, where there is still persistent ice. In addition, scientists have discovered that polar bears are mating with grizzly bears (Barnosky 2009). These changes in polar bear behavior are consequences of the secondary nature they face in a changing Artic ecology.

Changes in the Arctic ice ecology will also lead to further economic exploitation of the coastal and marine areas of the region. As the United Nations Environmental Programme (2013, 5) reports:

> Warming and melting of the sea ice and land snow offers greater human access to the Arctic region. Limited offshore oil, gas and mineral exploitation is already underway and will certainly increase in coming years, bringing new opportunities as well as increased risks of oil spills and pollution. Summer shipping through the Northern Sea Route along the Russian northern coast is beginning to increase, and traffic through the Northwest Passage is expected to grow, as is tourism and marine transport of goods. Some Arctic marine fisheries will become more accessible to regional and foreign fishing fleets.

While climate change will contribute to serious ecological and cultural impacts, economic growth and development in some areas will contribute to the globalization process.

Historically, indigenous peoples adapted to and responded to climate-related events (Barnes and Dove 2015). Cultural adaptation has long been one product of the coevolutionary relationships that exist between diverse peoples, places, and regions (Costanza, Graumich, and Steffens 2006). Cultures have adapted or failed to adapt to food and water insecurity brought on by climate change (Barnes and Dove 2015). Cultural adaptation requires a knowledge of place that is gained through direct human participation in nature. In time, experiential, intuitive forms of knowledge are developed and serve adaptation. These knowledge systems are based on a unique and intimate understanding of an ecology of place.

There is no guarantee that a culture will adapt to change, and there are many examples of cultures that failed to respond to changes in their ecological conditions. In many cases, adaptation required a deep place-based knowledge system that took thousands of years to develop. In *Collapse: How Societies Choose to Fail or Succeed* (2005), Jared Diamond describes five key factors that contribute to cultural collapse: climate change, hostile neighbors, collapse of essential trading partners, ecological problems, and failure to adapt to these ecological threats. Diamond describes several serious ecological threats that jeopardize our capacity to adapt: habitat destruction, loss of soil, drought, overuse of resources, the introduction of invasive species, pollution, energy shortages, and climate change. Whether we can adapt to the challenges we face will essentially be based on choices we make as individuals and as members of institutions. This places the burden of responsibility on each of us and on our respective communities. It is not merely a question of relying on government to respond.

The choice to continue to whittle away at nature's substance and integrity (the metaphorical stick) threatens our shared capacity to adapt. My fear is that our capacity to adapt and respond to the social and ecological changes brought on by the overuse of resources and climate disturbance is being diminished. Ecological insecurity is the direct result of turning away from the places we once depended on for nourishment and sustenance. We have little control over our own destiny—our water is transported from thousands of miles away, our food is imported by container ships that travel thousands of miles across the ocean, and our energy is derived from sources well beyond our horizon. As a consequence, we are increasingly vulnerable and at risk from the changes that are likely to occur in water, food, and energy availability. Substantial declines in water, food, and energy resources will result from climate change. Our dependence on the global economy leaves us increasingly at risk from climate-related changes to ecosystem services.

The scale of climate-related impacts has a tragic dimension. We live under threat of tragedy. The tragedy endangers what we hold in common and is produced by how we act in common. Everything is fundamentally at stake. If we are to adapt and respond to both the overuse of resources and the multiple impacts of climate change we will need to face tragic choices—choices about how we consume resources, how we treat and relate to nature, how we treat one another, and how we begin to restore our relationship to nature and community. Calabresi and Bobbitt (1978) describe the inevitability of tragic choices in modern institutions. Basic to the tragic form is the recognition of the inevitability of unresolved tensions that exist between diverse interests, beliefs, and values about government, nature, society, and our economy. In this case, the continued growth orientation of modern society is in conflict with the basic life-giving values associated with the planet. Simply reducing greenhouse gas emissions, for example, without addressing the deeper cultural (e.g., economic) and biophysical consequences of globalization is a tragic choice that contributes to ecological injury and social degradation. The question is whether human beings and their institutions are willing and able to make the tragic choice to protect and sustain the life-giving values carried by nature. If so, we will need to change the way we use resources, and protect enough of the remnant nature that exists to forge a future that can maintain the life-producing values of the biosphere. Industrial society lacks the ecological and communal sensibilities that are needed to respond and adapt to the substantive loss of ecosystem health and integrity. As consumers in a global economy, we remain disconnected from the place we inhabit, and are less aware of our surroundings. We are less capable of noticing the changes that we are causing. At diverse scales of social interaction we avoid taking the necessary steps to change and respond to risks and vulnerabilities. At stake is the diversity of the planet's life-support mechanisms, and the likely diminishment of the planet's cultural diversity.

At the same time that we are losing ecosystems we are losing the traditional cultures and their knowledge and ways (Maffi 2008; Maffi and Woodley 2010; Berkes 2012). In this sense there are biocultural impacts from climate change and overuse of resources. First, climate change has impacts on the biosphere by changing the life-giving characteristics of the planet. Second, climate change erodes the diversity of language and knowledge systems associated with the ethnosphere (e.g., the ethnic diversity of the biosphere). Ecological collapse is proceeded by the loss of a profound knowledge base that once served adaptation and cultural resilience. As we lose traditional cultures and their knowledge systems, we lose the capacity to learn from them.

SOMETHING WICKED THIS WAY COMES

One answer to the crisis is to turn to one's home place. A necessary first step to address the range of social and ecological threats and impacts we are facing is for individuals and communities to become more familiar with the places they inhabit. We need to cultivate an intuitive and experiential knowledge of our respective place, bioregion, or community.

A local farmer and I are walking his walnut orchard. He tells me that this is the first year in four generations that the walnut trees have not borne fruit. His cattle no longer have feed, and the grass is dry. The river has vanished. He has had to sell off his herd. We sit under a great old oak tree, and look across the Santa Ynez River valley. The branches of the walnut trees are blowing in the wind, and in the distance the fingers of a distant fog are spread out across the valley and foothills. There is a chill in the wind. He wonders out loud. He describes the impact of climate change on his crop and cattle, and wonders if his family ranch can survive.

A similar story is told by a fisher. His nets are empty. The sardines off the coast of California have crashed, and the federal government has closed the fishery this year. He tells me stories of fishing with his father for giant black sea bass and swordfish. Swordfish were harpooned while they slept on the surface of the sea. But now the big fish are gone. In their place, new marine resources are caught and exported. We have been fishing down the food chain for fifty years.

Ray Bradbury's *Something Wicked This Way Comes* (1962) is a novel that reveals the conflicting nature of good and evil that exists in individuals and society. Climate change is the wicked and dark by-product of our thirst for the "black gold" and coal used to feed industrialization and economic growth. For some, this dark side of modern civilization and our continued dependence on fossil fuels will culminate in a social, economic, and political diaspora, an unraveling of industrial civilization and biospheric destruction. Rather than perpetuating the denial of the inevitable collapse, the Dark Mountain Project is a network of writers, artists, and philosophers who are committed to reflecting the ecological reality of this diaspora. The project grew out of the despair and the belief that the humanities were failing to be honest about the scale of the impacts of wicked climate change. The members of the project hope that by writing and creating art they can offer a way of healing to establish a new foundation for changing the world.

This chapter represents a more hopeful response. There is a light emanating from the hearth and firepit. The darkness exists beyond the hearth, and is worth recognizing. The darkness exists on the horizon. But closer to home and in the warmth of the firepit we can find a hope and avoid the fear of that darkness.

There is no "silver bullet" that can resolve the climate crisis. Climate adaptation will require a variety of responses at diverse scales, and across different locations and regions. Richard Lazarus (2009, 1159) refers to the challenge posed by climate change as a "super wicked problem" that "defies resolution because of the enormous interdependencies, uncertainties, circularities, and conflicting stakeholders implicated by any effort to develop a solution." The super wicked nature of the multiple threats posed by climate change has to do, in part, with the complexity associated with the multiple scales of the impacts of the changing climate. While we are members of particular places we are also dependent on the life-giving values of the biosphere. The biophysical scale of climate change has local, regional, and global characteristics.

The wickedness of climate change has impacts on local ecosystems and the global biosphere. In this sense, the global impacts of climate change can vary from one region to another, and it can be difficult to predict local impacts on species diversity, habitats, and ecosystems. The diverse scales in the ecology of the planet's biosphere are analogous to the notion of the Sri Yantra. The *yantra* is a ritual object of Nepal, which represents the nucleus of the visible, and knowable, a linked diagram of lines that reflect particular energy sources. There are different kinds of *yantra*s, such as the Sri Yantra or Great Yantra. Other lesser *yantra*s (Om Yantra, Kali Yantra) are segments of the great embracing Sri Yantra. The notion of *yantra* serves as an analogy for the substance and energy of earth—the source of life, the connecting energy source that unites all earthly entities, including places and the people who inhabit ecosystems (see figure 1.1). It is this maintenance of the energy of the Sri Yantra or Planet Earth that is at stake today. The life-giving values of ecosystems produce our water, energy, food, and the air we breathe. We remain dependent on healthy ecosystems to survive. We also inhabit many of the lesser *yantra*s of earth. The sea can be considered a lesser *yantra* of the earth, and most of us depend on it for the protein it produces. The ocean is also a major contributor to the oxygen we breathe, and stores the carbon emissions from our burning of fossil fuels. A creek or river can be considered a lesser *yantra* of an entire watershed or river basin, with its tributaries linked to the sea. Water is the source of all life. Our hydrological modifications, pollution, overuse, and degradation of watershed ecosystems also contribute to our rising ecological insecurities. Each animal and habitat is connected to the greater *yantra* of the biosphere. The salmon swims upstream and downstream.

FIGURE 1.1 Sri Yantra of language, place, and knowledge.

These ecological connections are also reflected in the diversity of and interdependence of cultures and societies. Each aspect of the ethnosphere, as reflected in language and knowledge, can be considered a part of this biospheric *yantra*. Languages and knowledge systems are derived from generations of living in a particular place, as the stories are passed on from one generation to another, and as life's lessons are taught, remembered, and retold. Across the ethnosphere, communities are based on the intergenerational development of place-based language and knowledge that connects human beings to one another and to the natural world.

Over the past century and a half human activity has pushed the earth into a critical mode; four of the nine "planetary boundaries" have been crossed (Steffen et al. 2015). Biodiversity loss, fertilizer use, climate change, and land use are key planetary boundaries that have been crossed by human activity. A *tipping point* is the estimated point where an essential component of the planet's ecosystem can no longer function in the same way, nor can the system provide the types of ecosystem services that human beings depend on. Biodiversity loss is an important facet of the decline in the integrity of the biosphere. We are reaching the boundary of many of these biospheric tipping points, such as the substantive decline in native species diversity, and increasing social and economic risks and ecological insecurities will likely result.

One planetary boundary is the loss of biospheric integrity. This is the core of the entire planet's ecology, and will have cultural and social impacts. With respect to the warming of the globe, Steffen et al. (2015) write that a rise of 2 °C is a "risky target for humanity." An earlier study led by Hansen (2005) found that a warming of more than 1 °C, relative to 2000, will constitute dangerous climate change as judged from likely effects on sea level and extermination of species. Accordingly, Steffen et al. recommend a target closer to 1 °C in order to maintain both the climate and biospheric aspects of the planet's ecosystems.

We can expect substantive declines in primary and secondary levels of productivity of the world's ocean (Schubert et al. 2006). As the health of the ocean declines, the protein available from the sea declines. We can expect a major protein deficit in the near future; demand for protein sources from the sea increases, while the supply diminishes. A similar scenario is projected for cereals, grains, and other carbohydrates that are derived from farming. Supplies will be threatened by climate change, yet the demand from a growing population will continue to rise. All life will find it more difficult to adapt to the challenges that lie ahead. Scientific information clearly shows that we are losing essential terrestrial and aquatic ecosystems (Barnosky 2008; Steffen et al. 2015) and that significant degradation of the ocean's life-giving qualities is likely (Blunden and Arndt 2015; Baugrand et al. 2015; McCauley et al. 2015). In the journal *Science*, McCauley et al. (2015) indicate that marine ecosystem loss and degradation will increasingly become a major threat to the health and integrity of the biosphere. Climate change will impair the capacity of marine life to adapt to the other human impacts on coastal and marine ecosystems, and threats such as the continued warming of the ocean, sea level rise, and ocean acidification are likely unstoppable (Blunden and Arndt 2015). The major changes in ocean ecology will be longer-term. At least 1,141 of the 5,487 mammal species on earth are known to be threatened with extinction. One in four marine mammal species may go extinct. Between 1970 and 2010, the World Wildlife Fund (2014) reports the following global trends:

- Terrestrial wildlife is estimated to have declined by 39 percent.
- Marine life is estimated to have declined by 39 percent.
- Freshwater wildlife is estimated to have declined by 76 percent.
- Human population grew by 185 percent.

Simply reducing greenhouse gas emissions, for example, will not resolve the multiple problems and large-scale effects of ocean acidification (Feeley et al. 2008), or the steady decline in native-species diversity. There is no simple solution to the loss of endemic plants and animals that is caused by human activities. There is no simple institutional accommodation to the multiple threats and impacts associated with large-scale climate and ecosystem-based disturbance. In addition, we need to protect and preserve the traditional knowledge systems of the planet, because this knowledge has shown to be essential to cultural adaptation and resilience (Maffi and Woodley 2010).

Large-scale, centralized, and bureaucratic institutions are ill suited to address and respond to the social and ecological challenges associated with wicked climate change (Lazarus 2009). As Lazarus notes, "Ecological injury resists narrow redress; due to the highly interrelated nature of the ecosystem, it is almost always a mistake to suppose that one can isolate a single, discrete cause as the source of an environmental problem. A broader overview that accounts for the full spatial and temporal dimensions of the matter is needed. Failure to pursue such an overview is likely to result in an approach that is at best ineffective and at worst unwittingly destructive because of unanticipated consequences" (1181).

Lazarus provides a detailed characterization of why institutional reform and innovation is very difficult in the United States, given the intergovernmental framework and the fragmented nature of government authority. One reason for the failure of governmental response is the nature of intergovernmental conflict that is produced in climate-related adaptation and mitigation planning. In general, the larger the biophysical scale of the threat and pressure, the larger the scope of conflict between diverse interests, values, and belief structures in institutional processes. *Scope of conflict* is a term Schattschneider (1960) developed to explain how institutions often fail to address highly contentious issues and often respond by trying to control conflict rather than to resolve conflict. The scope of conflict is a reflection of the number of diverse participants in a decision-making situation. Schattschneider argued that government often attempts to control conflict by limiting the range of diverse voices, values, and interests expressed in an institutional context.

The importance of the connection between conflict and choice is described in Schattschneider's classic work, *The Semi-Sovereign People* (1960, 18): "There is nothing intrinsically good or bad about any given scope of conflict. Whether a large conflict is better than a small conflict depends on what the conflict is about and what people want to accomplish. *A change of scope makes possible a new pattern of competition, a new balance of forces and a new result,* but it also makes impossible a lot of other things" (emphasis added). As the scope of conflict expands, institutions will attempt to control the conflict by reducing the scale under consideration in the negotiating process. With respect to the scope of conflict and the large-scale characteristics of climate change, this process of negotiating ecology can be considered as follows. First, value-based conflicts emerge and are shaped by the physical or "characteristic" scale of the level of risks and impacts to ecosystems and society. The larger the scale of impact, the greater the conflict between values, interests, and beliefs in society. The more conflict in society, the less likely it is that decision-makers and stakeholders will support the large-scale institutional responses that are needed to address the myriad threats and impacts from climate change. The more conflict, the more likely that major policy development in support of climate adaptation will be put off to the future. For instance, climate adaptation and mitigation plans from government continue to emphasize the reduction of greenhouse gases and renewable energy development. Yet, reducing greenhouse gas emissions and renewable energy development cannot address the long-term social and ecological impacts of climate change. In this sense, reducing greenhouse gas emissions is an example of reducing the scope of the wicked problem to that of an energy or emissions issue.

In super wicked problems we cannot depend on scientific consensus to resolve conflict. Indeed, scientific information can contribute to the conflict over how to address and respond to climate change. One challenge is that there is less predictability in climate science as the scale of the problem and threat expands to consider larger-scale threats and risks. There remains a paucity of information on the social and ecological threats and pressures associated with climate change, and this is particularly the case when one considers an entire oceanographic province or biome. The argument that we should accord deference to the sciences in adaptation planning and decision-making arises from the ability of scientists to advance claims about consequences with a high degree of confidence. With respect to climate change, scientific confidence is low (or uncertainty is high). Given the high degree of scientific uncertainty, scientists lose their claim to special political deference in debates over why or how to respond and adapt to climate threats and impacts. In these decision-making situations, scientific knowledge is much less likely to be regarded as a reliable guide to the evaluation of loss or gain associated with particular consequences, such as the estimated loss of habitat or impacts on the integrity of a marine ecosystem.

The process of scientific discovery is not based on some type of general consensus among those who study nature. We cannot depend or rely on scientific consensus before we act. With respect to ecology as a science, Shrader-Frechette and McCoy (1993) show that ecology is fraught with basic uncertainties about issues central to the protection of native species and ecosystems. There are disagreements over methodology and epistemology that drive the scientific enterprise. The burden of proof, values, politics, and other qualitative aspects of decision-making play an influential role in the negotiation over how to address and respond to climate change. These disagreements and conflicts over how to respond are inevitable consequences of the intermingling of scientific facts and values.

Simply put, for the super wicked problem of global climate change (see box) there is no simple cultural resolution, legal remedy, institutional innovation, or social response that can address the multi-scale and multidimensional nature of the challenge. Building on this concept of the super wicked nature of climate change, Levin et al. (2012) define super wicked problems as having the following characteristics:

- Time is running out. The problem of climate change become more acute and difficult to resolve over time and across space.
- No central authority. There is a lack of policy and planning across scales to address ecosystem-wide and socio-cultural tipping points associated with climate change.
- Those seeking to solve the problem are also causing it. Our reliance on fossil fuels in the global economy continues to exacerbate the problem of global climate change. Government is unwilling to address and respond to climate change because of the perceived impacts on the global economy. While government focuses and defines the problems in terms of renewable energy use or greenhouse gas emissions, other issues related to climate change, such as biodiversity loss and the degradation of ecosystems, are not addressed by elected officials or by resource agencies.
- Policies discount the future irrationally. Issues of equity and fairness are deferred to future generations and those who are less capable of addressing and responding to the problem.

Properties of Super Wicked Climate Problems

- *The scope of change.* The more greenhouse gases continue to increase, the more dramatic will be the ecological and cultural consequences of climate change. The longer it takes to address and respond to the challenges of climate change, the harder it will be to maintain the ecosystems of the planet.
- *The challenge of adaptation.* Those who are more able to address and respond to the problems of climate change (e.g., developed countries) continue to cause more impact and have failed to develop plans to adapt to large-scale ecosystem degradation. There are few incentives in industrialized economies to curb the continued use of fossil fuels and to limit growth.
- *The scope of the problem.* No existing institutional frameworks and arrangements have the ability to develop an integrative, holistic, and comprehensive governance system that can address and respond to the multiple threats and impacts from climate change. Institutions rarely address the large-scale temporal and spatial features and properties of ecosystem-wide changes. There are no global-scale institutions to address and respond to the global-scale changes associated with climate-related threats and pressures.

Ultimately, adapting and responding to climate impacts necessitates ethical and economic choices about how to maintain security and how to resolve conflict (Barnett 2003; Barnett and Adger 2007). As a super wicked problem, the challenge of adapting to climate change inevitably involves a fundamental social, economic, and cultural paradox—we support the values associated with growth, and this growth is based on an addiction to oil and coal that significantly degrades and threatens the life-giving values of the biosphere. We continue to whittle away at the ecosystems we depend on for survival.

We can hope for large-scale changes and policy innovation by centralized government elites or international agreements or conventions. But there are practical and necessary small steps that should be taken at local and bioregional levels that can foster longer-term changes in our behavior and lifestyles. Wendell Berry (2015) writes:

> The needed policy changes, though addressed to present evils, wait upon the future, and so are presently non-existent. But changes in principle can be made now, by so few as just one of us. Changes in principle, carried into practice, are necessarily small changes made at home by one of us or a few

of us. Innumerable small solutions emerge as the changed principles are adapted to unique lives in unique small places. Such small solutions do not wait upon the future. Insofar as they are possible now, exist now, are actual and exemplary now, they give hope.

We can begin to address and respond to the impacts of climate change at the bioregional scale. I believe bioregional responses can be a hopeful turn to take responsibility for one's actions at a community-wide level; a community-based response can be the hopeful tone of light that is born out of the darkness and despair. Place remains a binding force, and a galvanizing force for change.

BIOREGIONAL ADAPTATION

A bioregional approach to adapt and respond to climate change can be an essential first step. A bioregional approach to climate adaptation can address issues that contribute to climate change and ecosystem decline, such as changes in local land use, regional biodiversity conservation, small-scale agricultural and fishing practices, and other human activities. A turn to greater understanding and strengthening of one's relationship to place can represent a first step toward addressing the range of local social, economic, and ecological challenges we face.

Bioregionalism is not another form of "environmentalism." An environment is something to behold, as if it exists outside of the human experience. The concept of "the environment" is a shallow one, devoid of the intimate relationships and partnerships that exist in diverse cultural systems. Ecological thinking reveals a much deeper approach to nature; the science and sensibility of ecology can provide us with way forward to strengthen our understanding and relationship with nature as a *life place*.

There are a number of articles and books on the subject of bioregional theory and practice (Sale 1985; Evanoff 2011; Lynch, Glotfelty, and Armbruster 2012; Cato 2012). I have provided a general survey of this literature elsewhere (McGinnis 1999a). Cultural historian Kirkpatrick Sale (1985) traced the evolution of bioregional theory to a two-hundred-year tradition of countercultural values that share a more critical stand against bureaucratic authority, centralized governance, and materialistic society. It can also be traced to the community-based lifestyles that exist across the world—from indigenous and tribal societies to contemporary provincial peoples. The diverse place-based bioregional movement can also be understood s a form of post-primitivism insofar as it is based on an ecological identity that recognizes the value of being "native" to a place. Becoming native to a place can only be gained by careful observation of walking a watershed, upriver and downriver, into the bush or forest, to the peaks of a coastal range, and then sharing the knowledge with others. It takes a long time to be a true resident and inhabitant of a place, and it takes cooperation and hard work with the other members of a community to be a citizen of an ecosystem.

Thomas Berry (1988) describes six "functional" characteristics of bioregional practice: self-propagation, self-nourishment, self-governance, self-education, self-healing, and self-fulfilling activities associated with place-based lifestyles and community-based identification. For Berry, the "self" is part of a more-than-human community: there is no separate world. A sense of an ecological self emerges in the practice of place (Thayer 2003; Berg 2015). While place can carry the resources that are needed for human beings and other species to survive, a place is much more than a bundle of commodities to be packaged, like lettuce or sardines, to be exported overseas. A place is a living community. As Robert Thayer (2003, 6) writes: "Embedded in the bioregional idea, therefore, is a very general hypothesis: that a mutually sustainable future for humans, other life-forms, and earthly systems can best be achieved by means of a spatial framework in which people live as rooted, active, participating members of a reasonably scaled, naturally bounded, ecologically defined territory, or life place."

One early advocate and activist of bioregional lifestyles was Peter Berg, who wrote, "You are in a bioregion, an ecological home place that has distinct continuities that affect the way you live and are affected by you" (quoted in Glotfelty and Quesnel 2015, 4). Gary Snyder (1990, 43), one of the great ecologically grounded poets living today, expresses a bioregionally oriented sentiment that clearly notes the difference between relying on centralized government and a place-based manifesto: "We seek the balance between cosmopolitan pluralism and deep local consciousness. We are asking how the whole human race can regain self-determination in place after centuries of having been disenfranchised by hierarchy and/or centralized power. Do not confuse this exercise with 'nationalism,' which is exactly the opposite, the impostor, the puppet of the State, the grinning ghost of the lost community."

BIOCULTURAL RESILIENCE

There are a number of common dimensions to the concept of bioregionalism and the more recent notion of *bioculturalism*. An integrative and holistic approach to viewing the world culturally and ecologically is through the lens of biocultural diversity (Maffi 1998). Bioculturalism recognizes the interdependence between cultural knowledge systems and language and the maintenance of ecological diversity (Hong 2013). Wade Davis describes the importance of language and diverse knowledge systems throughout the natural history of coevolution between peoples and places. In his 2007 TED talk, Davis stated, "Together the myriad cultures of the world make up a web of spiritual life and cultural life that envelops the planet, and is as important to the well-being of the planet as indeed is the biological web of life that you know as the biosphere." The ethnosphere comprises a diversity of place-based knowledge systems and languages. The relationship between situated knowledge, language, and place is a product of the evolutionary response of people who have adapted to the specific characteristics of a region's ecology. The coevolution of place and people is reflected in the language and knowledge systems that have been developed to respond to changes in bioregions (Maffi 1998). The synergy between social and ecological systems forms an experiential form of knowledge that fosters resilience and adaptation to changes that have often included major climate-related changes to regional ecosystems.

Maffi describes the importance of bioculturalism—the irrevocable connection between local peoples and their natural world and place—to cultural adaptation and survival. Place-based knowledge has long served cultural adaptation. Comprising about 4 percent of the world's population, traditional societies remain essential stewards of over 20 percent of the Earth's terrestrial and aquatic ecosystems, and therefore maintain roughly 80 percent of the native species diversity on the planet (Maffi 1998). The challenge today is to develop a place-based science and sensibility that can serve ecological resilience and adaptation.

The brutal facts of climate change have yet to sink in, and our adaptive capacity and capability are being put to the test. But more troublesome is the lack of intuitive and experiential knowledge of place. Accordingly, our choices of how to respond and adapt to climate change become shallower and less meaningful. As our collective responses to the multiple threats, pressures, and impacts of climate change are put off, the challenges of adaptation grow greater, our ecological insecurities become more pronounced, and our general capacity to adapt in time becomes less likely.

CALIFORNIA'S REVOLUTIONARY CLIMATE

California is one of the five Mediterranean-type ecosystems (MTEs) that have a rich natural history that includes long periods of ecosystem and climate-related disturbance events (Klausmeyer and Shaw 2009). Throughout the diverse histories of MTE-based cultures there have been many examples of adaptation and failed adaptation to climate change (Rundel, Montenegro, and Jaksic 1998). MTEs are far from homeostatic or stable systems (Blondel and Aronson 1999). The MTEs of the world are unique biomes that share a common natural history—human beings have had to adapt to major climate events such as flooding, earthquakes, fire, and changes in the availability of water and food (Barnes and Dove 2015). Indigenous societies in California adapted to climate change by migrating to other areas, changing their diet, and changing their behavior (Raab and Jones 2005).

This history of climate change is reflected in scientific studies in the Santa Barbara Basin that examine samples of sediment cores from the bottom of the Santa Barbara Channel. For the past twenty years scientists have been assessing sediment cores, and they find a history of oceanographic regime shifts that are both long-term and short-term climate events (Kennett and Ingram 1995). Long-term changes in sea surface temperature, or what oceanographers refer to as *regime shifts*, reflect consistent transitions from warmer to cooler water and back again. For thousands of years, marine life adapted to these shifts by moving north, up the coast, to the cooler waters of the California Current (Moffitt et al. 2015). There is no evidence that marine life became extinct during the last several thousands of years of long-term shifts in sea surface temperature. Shorter-term shifts in sea surface temperature are associated with the cycles of El Niño (warmer water) and La Niña (cooler water). Humans and other species coevolved to adapt to these colder and warmer water regimes. But because of human use of marine species and the destruction of habitat, both on land and in the sea, species like the white abalone were not able to adapt.

At the regional scale, the diverse ecosystems of California are influenced by climate change in many different ways. Changes in the oceanographic currents, the atmosphere, and biology influence the ecosystems of the state. In general, California is a "revolutionary" climate that includes a natural history of long-term droughts, major flood events, earthquakes, and fire. California experienced long-term droughts or extreme hydrological shifts in 892–1112 (220 years) and 1209–1350 (141 years) (Davis 1991). The longest drought of the twentieth century lasted only six years, during 1987–1992 (Priest 1993). Today's drought is the worst in 1,200 years (Griffin and Anchukaitis 2014).

The California Floristic Province is one of the most important areas for biodiversity in the world. Like other inhabitants of Mediterranean-type bioregions, Californians are increasingly vulnerable to shortages of water, food security issues, changes in climate, and other potential threats and pressures. In virtually every place you walk, the history of a changing landscape is under your feet. Climate change will lead to dramatic loss of native species diversity. Many species that are listed as threatened or endangered due to human impacts, such as habitat fragmentation, overuse, and the introduction of invasive species, will find it more difficult to adapt to climate-related impacts. Climate change will impair the capacity of species to adapt.

One consequence of climate disturbance in California will be a shift of biodiversity to the north (Loarie et al. 2008). The native plants unique to California are very vulnerable to global climate change, such that two-thirds of these "endemics" could lose more than 80 percent of their geographic range by the end of the century (Loarie et al. 2008). Scientists at the U.S. Geological Survey developed the Coastal Vulnerability Index to assess the physical vulnerability of the California coast. They found that from San Luis Obispo to the Mexico border, communities have "high" or "very high" vulnerability to climate change. Many areas in the coastal watersheds of California are recognized as threatened "hot spots" for biodiversity (Stein, Kutner, and Adams 2000). Climate change has direct and indirect impacts on these species and their habitats, especially designated critical habitat and environmental sensitive habitat areas. Existing protected areas, such as ecological reserves, wildlife areas, mitigation sites, and easements will likely be impacted by climate change.

Indigenous cultures in California also adapted to short- and long-term changes in the climate. Anthropologists have shown that before the colonization, terror, and conquest of prehistoric California early tribal societies adapted to climate change in myriad ways (Raab and Jones 2004). Like other tribal societies, prehistoric societies in California adopted to the drought and famine that were often associated with long-term and large-scale climatic events (Barnes and Dove 2015). These resilient indigenous societies were irrevocably connected to the landscapes and seascapes they inhabited, and their biocultural knowledge served adaptation across thousands of years. Biocultural knowledge was grounded in an intuitive sense of the changes that were occurring in their respective bioregions, and the knowledge gained by direct human participation in a more-than-human community.

The historical record shows that there have been long periods of drought, famine, fire, and other major disturbance events in California. For thousands of years indigenous peoples adapted to these events by developing new social norms and behaviors (Fagan 2004). In many cases, adaptation contributed to migration of traditional peoples to safe havens or refuge areas (Barnes and Dove 2015). In the case of California tribal peoples this included migration from inland areas closer to the coast and river systems. Migration was not without conflict, and new archaeological studies show that tribes were in conflict over scarce resources during difficult times like climate events such as long-term drought, flooding, and famine (Raab and Jones 2004). California's indigenous cultures adapted to long-term climate events by using traditional ecological knowledge systems (Johnson 2000). The coastal inhabitation by early maritime cultures of south-central California reflected unique place-based languages and kinship relationships (McGinnis and Cordero 2004).

The languages and customs of diverse indigenous Californians often reflected the soft boundaries of a watershed, river basin, or bioregion. For example, the diverse Chumash people spoke different but related languages in distinct but interdependent parts of the bioregion. The people were heavily dependent on a healthy marine environment; the marine component of the Chumash diet consisted of over 150 types of marine fishes as well as a variety of shellfish including crabs, lobsters, mussels, abalone, clams, oysters, chitons, and other gastropods. Shellfish were essential to the Chumash economy and material culture (Erlandson et al. 2011). Archaeological research shows that Chumash people adapted to climate change by changing their use of coastal marine resources (King 1990).

Portions of coastal areas on the northern Channel Islands were sites of Chumash villages, and are now submerged by changes in sea level. Thousands of years ago the sea level was at least 150 feet lower than it is today, and the northern Channel Islands were joined as one island. Recently discovered paleontological remains have also contributed to the rich record of the coastal area. In 1994, for example, a relatively complete pygmy mammoth was discovered on a coastal bluff on the north shore of Santa Rosa Island. This discovery represents the most complete pygmy mammoth discovered in the world to date. Early human remains of a woman (named Arlington Springs Woman) were discovered at Arlington Canyon on Santa Rosa Island, dating back to the end of the Pleistocene, approximately 13,000 years ago.

It is difficult to imagine the depth of cultural values and understanding of place in prehistoric California that fostered adaptation. New cultural norms and behavior emerged to respond and adapt to major climate-related events in California (Raab and Jones 2004). The establishment of sacred lands was likely an important part of the preservation of refuge for species and unique plants that served adaptation. Descendants of the Chumash consider the northern Channel Islands a special place, still occasionally paddling these waters in *tomols* or wooden canoes (McGinnis and Cordero 2004). The first *tomol* to be owned by the Chumash people since the 1880s is the *'Elye'wun* (swordfish), which was built by the Chumash community in 1996–97 under the leadership of the Chumash Maritime Association. The swordfish is symbolized by abalone inlay carved and embedded in the bow of the *tomol*. In this sense, the *tomol* is the swordfish and leads the paddlers to the island. The building of the *'Elye'wun* and the crossing to Santa Cruz Island is a manifestation of a new effort by the Chumash people to reconnect and restore their relationship to the sea and the northern Channel Islands. The ceremonial paddle across the Santa Barbara Channel from the coastal mainland represents the culture's affirmation of the deep connection between the Santa Ynez Mountains, the coastal watersheds, the marine system, and the northern Channel Islands (map 1.1).

Abalone was a staple of the Chumash diet (Erlandson et al. 2011). White abalone (*Haliotis sorenseni*) is a marine snail, a deep-water species found at between 80 and 200 feet on rocky reefs from Point Conception to Punta Abreojos in Baja California, Mexico (Leet 2001). Highly prized for their tender white meat, white abalone were harvested in an intense commercial and recreational fishery that developed during the 1970s, then quickly peaked and crashed as the abalone became increasingly scarce. The biological collapse of white abalone was brought on by changes in sea surface temperature, overfishing by commercial fishers, and withering foot disease. As the abundance of the species declined, its capacity to adapt to the changes brought on by human beings was diminished. The rarity of this species within its historical center of abundance prompted the National Marine Fisheries Service to list it as a candidate species under the Endangered Species Act in 1997. In May 2001, the white abalone became the first marine inver-

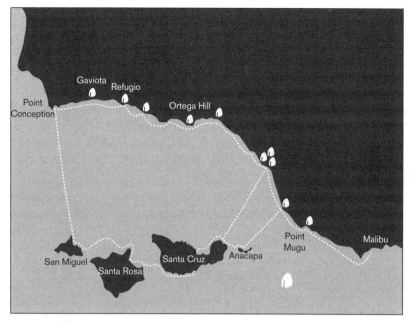

MAP 1.1 *Tomol* routes of the Chumash, off south-central California, in 13,000-plus years of inhabitation.

tebrate to receive federal protection as an endangered species in the bioregion (California Department of Fish and Game 2001). The plight of the white abalone is a symptom of a much larger-scale threat to the coastal marine ecosystems of the bioregion. Note that this was the first benthic (bottom-dwelling) species to be federally listed. The listing of such species as threatened is an indication that human activities challenge the adaptive capacity of many species that have historically been able to respond to climate-related events.

BIOREGIONAL INTEGRATION

One fundamental question is whether we have the necessary intuitive and local, place-based knowledge to recognize the multiple dimensions of threats posed by climate change. We need to cultivate a renewed sense of place and community as a first response to the challenge of maintaining biocultural systems in an age of climate change. Without a greater understanding of the particular ecosystems we inhabit, we cannot notice the change in the local ecosystems we depend on. A place-based approach to climate adaptation requires a more integrative, holistic, and ecologically grounded practice and lifestyle.

Unlike in the past, there are few safe refuge areas to migrate to in this Anthropocene age—we have diminished the life-producing qualities and characteristics of the biosphere to the point that there are few climate refugia remaining for either human beings or other life forms to adapt. There are three interrelated challenges today. First, we no longer have refuge areas to migrate to, to survive the consequences of climate change. Second, the consequences of today's social and ecological crisis brought on by climate change and the overexploitation of resources are much more complex and devastating than the past. Third, we lack a knowledge-based and intuitive understanding of the nature of the crisis we face, and are at present unable to respond and adapt to the wicked nature of climate change.

In response, the formidable task of beginning the long-term process of adaptation includes a fundamental economic challenge to resist the "grinning ghost" of globalization. First, working in the bioregional template requires changing our behavior to respect the productive capacity of local ecosystems. The trade of local-scale resources to global markets ultimately undermines the ecology of the planet. Second, because of the multiple values associated with ecosystems, there are many stakeholders, interests, and user groups that will be in conflict over the access to, use of, and distribution of resources. Consider those that are always present: governments and consumptive users of the sea. In addition, there are activists and would-be investors from outside. These groups have conflicting goals and values. The art of integrative, bioregional, place-based adaptation under these conditions is to integrate conflicting values, at least temporarily, so as to make agreements on resource use and other protective actions that are needed to identify and sustain essential climate refuge areas, and that can better reflect the productive capacity of local ecosystems. Third, there are the challenges of adaptation and resilience, which require new modes of learning and partnership building. We need to stop whittling away at the stick, protect essential ecosystems to serve ecological resilience, and curb the overuse and globalization of resources. Local ecosystems cannot be sustained if they remain traded as commodities in a placeless global market.

To strengthen the localization process of economic production and consumption, figure 1.2 includes a depiction of three key values of bioregional integration. Bioregional integration (see box) includes the value of protecting a people's connection and relationship to place, their

Ecology

Multiscalar System
Governance

Ecological
Sustainability

The Maintenance
of Ecological
Integrity & Health

Place-Based
Economic
Development

Face-to-Face
Democracy

Equitable
Allocation
of Public
Goods

The Promise
of Civil Society

Equity

Economy

FIGURE 1.2 Three key values of bioregional integration.

specific landscape or seascape (water, food, and air—or collectively, the "commons").

Science and scientists have a role in informing society, but the quest for a "recovery of the commons" (Snyder 1995a) will require a fundamental shift in value orientation and, in some circumstances, the protection of the cultural values that support the "natural contract." We cannot predict or control ecosystems; we can only control human behavior (e.g., our modes of production and consumption) and associated impacts. We influence ecosystem dynamics, but we cannot manage ecosystems.

THE PATH FORWARD

In *Space and Place: The Perspective of Experience* (2001), geographer Yi-Fu Tuan puts forth the equation, "space plus culture equals place." Tuan's equation can appropriately be applied to the landscapes and seascapes we depend on and are irrevocably connected to. Global climate

Principles of Bioregional Integration

A number of principles support bioregional integration, including but not limited to:

- restoring the relationship between place and the economic mode of production and consumption
- protecting rural lands and local agriculture
- establishing biodiversity policies that also support the values of inter-generational equity and environmental justice
- developing regional markets for regionally produced products
- creating value-added programs for sustainably produced resources
- assessing and monitoring economic strategies based on their ability to detect long-term trends in changes to economic security, and with respect to the causes of significant ecosystem change
- emphasizing cooperative, inter-jurisdictional, cross-boundary conservation partnerships that can cultivate the necessary place-based, social, and economic alliances, with potential new roles for government and nongovernment groups.

change has a large-scale spatial and temporal dimensions that are difficult to comprehend. A space has little soul or spirit apart from the interdependent natural features of the places we occupy. Space becomes place and one's region when social identity is influenced and shaped by the vernacular and provincial characteristics of home and community. With respect to the impacts of climate change, we cannot fully understand place without acknowledging the impacts of climate disturbance on the remnant native features and on our respective communities. Space becomes place when we recognize that nature not only provides the sustenance of life but is also the source of ritual, myth, story, and cultural celebration. A return to one's place can also represent an initial step toward adapting and responding to the impacts of climate change on our communities.

Though we have been taught to observe the world objectively, far removed from a natural world, we need to learn to be embedded in the systems we reside in, and relearn the meaningfulness and mindfulness of inhabitation. We need to be more engaged, attached, aware, and attentive to place, not as mere observers of an "environment" or consumers of resources but as participants in the more-than-human community. This book calls for a "recovery of the commons," which requires a fundamental shift in values to primarily supporting the life-giving qualities of ecosystems, rather than the short-term values of maximizing financial return and global resource use. Gary Snyder (1995a, 36) recognizes that "we need to make a world-scale 'Natural Contract' with the oceans, the air, the birds in the sky. The challenge is to bring the whole victimized world of the 'common pool resource' into the Mind of the Commons. . . . There is no choice but to call for the recovery of the commons, and this in a modern world which doesn't quite realize what it has lost."

This book includes case studies that explore the role of science and sensibility in how we relate to a river system, watershed, and coastal marine areas in diverse areas across the Pacific Ocean—in California, New Zealand, and the South Pacific. These case studies were chosen because most of us live along a river or a coastal area and are dependent on the ocean for survival. Today, over 50 percent of the world's population lives within 3 km of a surface freshwater body, and only 10 percent lives further away than 10 km (Kummu et al. 2011). In addition, more than half the world's population lives within 60 km of the shoreline, and this could rise to three-quarters by the year 2020. This means that most people depend on coastal marine and/or river systems for their ecological security. As noted earlier, these ecosystems are some of the most threatened in the world (Finlayson, D'Cruz, and Davidson 2005).

I refer to the relationship between science and values as a political process of *negotiating ecology*. The chapters address a number of tenuous relationships that exist today in this Anthropocene age: watersheds and the plight of wild salmon; dairy production and the creation of *wastesheds*; offshore oil rigs and the ecology of fishes; container ships and the killing of whales; and sea level rise and displacement of island peoples, among other issues and concerns. The book also offers a number of responses, including: localizing the primary modes of economic consumption; the protection of marine life; watershed-based activism; and the restoration of the landscape and community.

There are three major parts to the book. The first part focuses on rivers, watersheds, and wastes. In this section, I explore the relationships between science and values in the California watershed movement, coastal adaptation in the islands of the South Pacific, and watershed planning in New Zealand. Part 2 explores the role of science and values in marine systems in offshore California, the Gulf of Mexico, and New Zealand. Part 3 includes chapters on the importance of place and community as an essential prerequisite for cultural adaptation and ecological resilience. The goal of the last section is to explore ways to recover a sense of place and community—or what Thoreau called in his essay "Economy," in *Walden* (1854), "the nourishing quality of the soil of the soul"—by describing how the cultivation of a place-based ecological sensibility represents an important first step in responding to the many challenges we face.

The case studies and respective chapters are reflections of a personal journey that I draw from as an educator, professional, and academic. One underlying theme is the importance of bioregional thinking and behavior. The book describes a number of principal elements of bioregionalism: homecoming, watershed-based activism, rethinking the mechanical sensibility, place-based consumption and preservation, respecting traditional knowledge, and bioregional restoration.

Homecoming

Chapter 2 is a reflection of my own homecoming and my place. An essential first step to addressing the particular problems we face in society and our treatment of nature is to become more acquainted with the other members of our more-than-human community. We are members of a community and bioregion that includes a circle of animals, plants, and insects. The word *community* derives from Latin *munus*, which has a number of meanings that are relevant here, including service, duty, gift, and sacrifice. A community is an assemblage of individuals bound by a relationship and partnership. This relationship and partnership are based on mutual obligation, exchange of gifts, and shared service. Homecoming is a process whereby one gains a deeper understanding and knowledge of one's place in the world and the needs of other species. I draw from my own experiential knowledge of my own place and region.

Watershed-Based Activism

Re-inhabiting a particular watershed commons represents an ideal place to start the homecoming process. Working with other members of a community is a trait of humanity, even though it seems to have been lost with respect to our globalizing modes of economic consumption and industrial production. Chapter 3 describes the evolution of watershed-based activism in California. A good way to start thinking about your relationship to the greater watershed is to engage and interact with it. Participate in the river's flow, swim upstream or downstream as if you were a salmon, and cultivate a sensory and olfactory memory of place as if your livelihood (and that of your family and neighborhood) depended on it.

Chapter 4 describes the decline of watershed health across New Zealand. Despite the country's "100% Pure" branding, the watersheds of New Zealand are showing significant signs of degradation. This is, in part, one consequence of a major change in how the land is used and abused in the country. The mechanization and industrialization of New Zealand's watersheds has led to the creation of wastesheds: the watershed becomes a source of source of pollution and wastes. I also explore the importance of cultural heritage and ecological identity as ways to better conserve and restore the watersheds of the country.

Rethinking the Mechanical Sensibility

The mechanization and industrialization of the land is reflected in our poor treatment of marine ecosystems and our continued reliance on fossil fuels. We need to think more critically about the myriad impacts of the mechanization of place, and the impacts of this dependence on fossil fuels on our particular places and regions. An ecologically literate community understands that marine systems are a source of life; they cannot be replicated or replaced. Chapter 5 explores the politics and ecology of the future decommissioning of offshore oil activities in the marine areas of California and the Gulf of Mexico. Hundreds of offshore structures will be decommissioned in the near future. Public policy in these two marine areas is driven by values and diverse interests in the future of these offshore structures, and perceptions of what is "natural" or "artificial."

Place-Based Consumption and Preservation

Chapters 6, 7, and 8 focus on the need to protect marine life across the Pacific Ocean. The human impacts on marine ecosystems are a sign of the Anthropocene age, which is an informal geologic chronological term that marks the impacts of human activities on the earth's ecosystems. The term was coined in the 1980s by ecologist Eugene Stoermer and popularized by a Nobel Prize–winning atmospheric chemist, Paul Crutzen. The oceans are threatened by a combination of two modes of globalization: the global impacts of climate disturbance and the global scale of economic consumption of marine resources. Coastal marine management should take into account the large-scale impacts of climate disturbance on coastal marine ecosystems that are being driven by the rise in sea surface temperature, changes in salinity, increasing acidification (changes in oceanic pH), and general declines in primary and secondary levels of productivity of the world's oceans (Schubert et al. 2006). Yet, as these chapters argue, human beings cannot control or manage the ecological processes that influence the productivity of the sea, we can only control human behavior, our consumption of marine life and associated impacts.

Without substantive change in value orientations and the approaches used to address the human impacts on coastal marine ecosystems, we will become more vulnerable and less secure in the face of climate change. Chapter 6 describes the politics and science that have influenced the designation of marine protected areas in California. Chapter 7 describes the challenge of marine governance in New Zealand. Chapter 8 describes the tenuous relationship between our reliance on container ships in our global economy and the whales of the California Current. Ultimately, the protection of the world's ocean requires the cultivation of place-based ocean constituencies that can support a renewed maritime sensibility and ethos—one that embraces not just the economic or instrumental use of marine resources that are traded in a global economy but recognition of the intrinsic place-based values of healthy coastal marine systems. Maritime communities should be recognized as distinct places worth protecting. Protecting maritime place requires the preservation of marine ecosystems. In a context of global climate change and global economy, a more ecologically integrative and holistic approach to resource allocation and biodiversity preservation is needed that can support particular maritime places and communities. Policies that support the up-scaling of marine life protection should be combined with down-scaling of the economic use of marine species in global markets to ensure that maritime communities can adapt and be sustained.

Respecting Traditional Knowledge

Chapter 9 focuses on the challenges faced by island peoples in this age of climate change and rising sea levels. With rising sea levels, islands are being lost and indigenous peoples forced to migrate, often to distant places. The climate refugee is placeless, a passenger on a sinking ship without a lifeboat. There is a need for a broader theory of justice that can embrace other forms of knowledge and life in an age of climate change, and one that can address the plight of climate refugees. With the loss of these cultures and communities, we also lose the place-based knowledge systems that have long played a key role in supporting adaptation to previous climate-related changes. I propose a theory of biocultural justice that respects the diversity of epistemologies of maritime place, and links alternative knowledge systems (both traditional forms of ecological knowledge and scientific knowledge) with principles of sustaining ecological security. Biocultural justice represents a shift from "shallow," anthropocentric theories of social justice to a "deep" practice of an ecologically based theory of justice. Schlosberg (2012) and others support a spatial expansion of an epistemology of justice, horizontally into a broader ecological range of social issues and vertically into examinations of the global nature of injustices that are associated with food, energy, and water insecurity. This conceptual shift underscores the need to support a theory of justice that represents a deeper realm of human relationship with the more-than-human world, where protecting marine ecosystem health and integrity are understood as essential principles and conditions that can support a practice of justice. A concerted effort at the international level is needed that can support a theory of blue justice. For those who receive the burden of the socio-ecological costs and risks from climate change, international agreements and conventions are needed that can encourage the protection of people who will need to be relocated in the future.

Bioregional Restoration

Restoring place is the focus of chapter 10. The imagination plays a critical role in the path toward bioregional restoration. Bioregional restoration requires not only the act of replanting native plants along a creek's bank, digging out and removing invasive species, or restoring a river's flow to bring back wild salmon. Bioregional restoration also includes storytelling, theater, art, and ceremony that strengthen the connection between people with place. For many years, I dressed up as a salmon and gave presentations to children about the values of salmon. I told stories of the salmons' migration upstream, and discussed the ecology of salmon along the south coast of California. We also discussed the differences between the role of native habitat in the salmons' return, and what role our community has in restoring wild salmon to our creeks and rivers. Wearing the mask of the salmon was one way to strengthen the connection to and awareness of the wonder of the natural world. The hope was that children would identify with the salmon, and perhaps teach their parents about how we can restore the species to our neighborhood creeks.

. . .

In the bright white and yellow glow of this day,
we turn to the mountains and think of the sea.
Today we can begin the hard work of community building—to
    celebrate our connection to a creek, to the greater watershed,
to the bioregion—to celebrate our gift from the blue islands on our
    horizon.
We are grateful for the gifts of this place and region.
It is the breath of our song, our clear winter breeze.
Let's celebrate this place with renewed partnerships, and a renewed
    sense of community and continuity with others.
We should be grateful to the wild beings, their secrets, freedoms,
    and ways.
We should be grateful to the sun-facing, light-changing sycamores.
We depend on this earthhousehold by the sea, this amphitheater by
    the sea.
There are shadows of wild southern steelhead in this creek.
We should cherish the springtime bloom of ceanothus and Chinese
    houses, monkey flower,
and the interface between the land, fresh and saltwater.

CHAPTER 2

Household Words

Cultivating an Ecological Sensibility

We aspire to live in the Household affections, and to be
numbered among the Household thoughts, of our readers. We
hope to be the comrade and friend of many thousands of
people, of both sexes, and of all ages and conditions, on
whose faces we may never look. We seek to bring to innumerable homes, from the stirring world around us, the knowledge
of many social wonders, good and evil, that are not calculated
to render any of us less ardently persevering in ourselves, less
faithful in the progress of mankind, less thankful for the
privilege of living in this summer-dawn of time.

—Charles Dickens, "A Preliminary Word" (1850)

"I would ask you to remember only this one thing," said
Badger. "The stories people tell have a way of taking care of
them. If stories come to you, care for them. And learn to give
them away where they are needed. Sometimes a person needs
a story more than food to stay alive. That is why we put
these stories in each other's memory. This is how people care
for themselves. One day you will be good story-tellers. Never
forget these obligations."

—Barry Lopez, *Crow and Weasel* (1990)

Homecoming is a process whereby one gains a deeper understanding
and knowledge of one's place in the world. This process can contribute
to our shared capacity to adapt, respond, and maintain the essential
characteristics of the regional ecosystems we inhabit. It is very difficult
to protect ecosystems without the sense of place and knowledge that

can be drawn from an understanding of the more-than-human features and ecological characteristics of a bioregion. Students of environmental studies are rarely introduced to the ecology of their region or their home. We often remain detached observers of the places even though we are dependent on the ecosystem services and goods (such as water and food) that our home place provides each of us. Psychologists refer to this detachment from the natural world as a form of *nature deficit disorder* that can contribute to a range of maladaptive social problems. In urban and rural settings, most people have limited knowledge or understanding of their immediate natural surroundings. Richard Louv (2005) finds that disengagement between human beings, especially children, and the lack of experience with the natural world has profound cultural implications not only for the health of future generations but for the health of the planet.

We don't know where our water comes from, where our food is produced, or where our wastewater and trash go. We are citizens of counties, states, and countries, and fail to consider a sense of citizenship that is attached to particular places. We consume goods and resources that are traded in a global economy with very little understanding of the ecological and cultural consequences of this mode of exchange. If we value the protection of society and nature we need to take the first step in re-inhabiting the regional ecosystems we benefit from on a daily basis. This process of re-inhabitation requires the use of scientific information and a renewed sense and ethos in support of place and community.

When you inhabit a place you recognize the patterns of the wind, the first flowers of spring, the variety of birdsong, and the paths of animals. These tastes, feelings, sights, sounds, and sensations enter the body and mind. But they must not be held in isolation. Feelings and sensations of place must be expressed and shared in a community. The communication of an ecology of specific areas is one role scientists can play in society. There is a role of storytelling and place-based narrative that can promote an ecological sensibility as well. With elements such as winds, trees, and rivers, this landscape should be celebrated as an essential part of a region's history and vernacular culture.

THINKING ECOLOGICALLY

The science of ecology provides a necessary foundation to explore one's place and regional home (Code 2006). Science is a way of thinking about the world. Scientific observation requires that our eyes be open, that we listen carefully, and that we notice the subtle changes in the landscape. We can trace the early use of the term *science* to the Latin *scientia* (knowledge) and the Ancient Greek *epistēme* (meaning). The older and closely related meaning of *science* as a form of knowledge and meaning in the world was closely linked to *philosophy*. In the early modern period the terms *science* and *philosophy of nature* were often used interchangeably. During the Scientific Revolution, natural philosophy (which today is called natural science) was considered a separate branch of philosophy. In many ways, the foundation of knowledge refers to the meanings one gives to a particular place or locale.

Ecology is based on the early Greek terms *oikos* and *logos,* the combination meaning "study of the house." The interconnections and processes that link people, nature, and place can be understood in terms of what the Ancient Greeks referred to as *physis* (φύσις), which is translated as the essential fullness and life-giving aspects of a living earth. A plant, for instance, sprouts from the soil, reaching to the sun, and this is the unfolding quality of life. The notion of physis also includes the withholding of life, and the processes that include, for instance, the death of a plant. The plant withdraws and returns to the soil. The plant's organic qualities become part of the natural cycle of growth and regeneration. The plant and soil are reunited. In his writings on nature, Aristotle uses the term physis as a source or cause of being—being born and dying. Aristotle contrasts natural things with the artificial. Mechanical or human-made artificial things move according to what they are made of, not according to what they are (Tarnas 1991). Aristotle's view of physis as the origin of life is directly opposed to the construction of nature as a machine.

Thoreau was a student of Greek philosophy. In *Walden*, physis is described (in the chapter "Spring") as the transition from darkness to light. Thoreau feared the rise of the machine in daily life insofar as he believed it would ultimately separate us from the natural world. Few people understand the linkages between mountains, rivers or creeks, coastal valleys, and marine systems. Human beings in an industrial society like ours are more attached to their computers and other technologies. We remain detached and alienated from the natural world, and have grown increasingly dependent on the technological machinery that has reshaped our communication and relationships with one another and with nature. We need to recognize that there is no separate reality that exists independent of the movement of waves, the direction of the winds, the quality of the soils, the flow of a river or creek, or the movement of animals. There is a close relationship between what it means to be human and the health of watersheds, plants, and animals—collectively, the bioregion.

A first step toward ecologically based awareness is to understand the characteristics and major features of the landscapes and communities that you inhabit. In *Bioregions: The Context for Reinhabiting the Earth,* Thomas Berry (1982, 5) writes: "The bioregion is the domestic setting of community just as the home is the domestic setting of the family. The community carries within itself not only the nourishing energies that are needed by each member of the community; it also contains within itself the special powers of regeneration." This special power of regeneration requires a particular place-based sensibility and a deeper ecological awareness of one's home. It also requires that we rethink the values and cultural worldviews that support technological progress, economic development, and the mechanization of society and nature.

As a student of ecology, I began a process of re-inhabitation by exploring the connections that I have with my place. This process can begin with becoming more aware of where your food comes from, the seasonal variability of different crops, how the produce you eat was farmed, and what types of impacts farming has had on the ecosystem. Even if you live in a city, you can learn that the system of shared services in an urban community connects culture to the natural and rural land-scapes. As author and activist Freeman House (1999, 99) notes, "There is no way to learn to live in place but from the place itself." Place is a living region that can foster a renewed sense of community that serves re-inhabitation. House also writes: "The history that will best serve the ends of learning to live as communities of place—to learn where we are—lies in the landscape surrounding us. . . . It is up to us who are alive now to translate this information into something like the place's own memory" (160).

A sense of place also requires recognition of the factors that contribute to changes in the landscape and seascape, including the threats and impacts from climate change. These changes can be as subtle as the slow change of sea surface temperature or as dramatic as the rapid decline in the abundance of songbirds during springtime.

NEGOTIATING AN ECOLOGY OF PLACE

To better adapt to the changes that lie ahead we need to begin a process of homecoming—one that combines the use of scientific information and a deeper sensibility of our place in the world. Negotiating an ecology of place is fundamental to the cultivation of a shared adaptive capacity. Anthropologists consider local knowledge key to the development of a culture's resilience in times of change. Recognizing the changes that are occurring in the natural world, for example, can strengthen people's ability to adapt to issues such as food and water insecurity. The presence or absence of a particular species is a harbinger of change. Such harbingers of change are all around us. You may witness the presence of humpback whales offshore. Whales are off the coast, feeding on abundant anchovies. This is part of a cycle that the careful observer of the marine system recognizes. But in the absence of local knowledge these signs of change are difficult to understand and respond to. Change is in the air. It is the context that we currently live in. Change is part of living-in-place; responding to change depends on the knowledge gained by participating in a more-than-human community.

Homecoming requires a recognition of what has been lost in the natural world, and how our modes of economic production and consumption are contributing to this change in the ecosystems we depend on. Changes in California's native species diversity and habitats are worth noting. In California, examples of aquatic ecosystems that have been largely eliminated from Southern California include: estuarine wetlands (i.e., salt marshes) as an entire subsystem, at 75–90 percent; the "riparian community," at 90–95 percent; and vernal pools, at 90 percent. Overall, 95 percent of the coastal wetlands of Southern California have been destroyed (California Coastal Conservancy, 2002). The losses of these habitats have contributed directly to the reduction in coastal and marine biodiversity of Southern California, as evidenced by estimates that 55 percent of the animals and 25 percent of the plants designated as threatened or endangered by the state depend on wetland habitats for their survival.

A study conducted by researchers at Stanford University's Center for Ocean Solutions (Caldwell 2009) included a review of over 3,400 peer-reviewed articles and an analysis of the primary threats to the Pacific Ocean. The study identified four primary threats described in the scientific literature—pollution, overfishing, habitat destruction, and climate change. With the development of new techniques, scientists are describing the synergistic impacts of multiple use and anthropogenic pressures on coastal marine systems. Human impacts (overfishing, pollution, and habitat degradation, among other pressures) and climate change impair an ecosystem's ability to withstand stress and associated disturbance events.

PLACE-BASED KNOWLEDGE

The biological kingdom, Animalia, is composed of many species, but
psychologically it is we who are composed of animals.
—Paul Shepard (1996, 89)

The relationship we cultivate with the landscape and the seascape is
shaped by local knowledge of particular places. Knowledge of the
unique features of the landscape takes time and commitment to learn—
the language of the birds, seasonal food production, when to fish, and
why we need to protect and preserve agriculture and the natural features of the landscape and marine habitats. As citizens of places, our
role is to understand the ways of the natural world we depend on, and
to garner support and collaboration to respond to these changing landscapes to ensure ecological resiliency for future generations.

Finding a place to dwell is analogous to the knowledge it takes to
cross the channel in the Chumash *tomol*. The *tomol* is built from redwoods washed up on shore. It is carved out and carries people across
the sea. The knowledge of building a seaworthy *tomol* is one key to
surviving the voyage. The knowledge of when to cross the channel is
another valuable insight. What is the direction of the wind? Is there a
storm approaching? Can you smell the rain on the horizon? What do
the animals tell us about the condition of the marine environment?
Once the *tomol* reaches the island, there is a celebration of homecoming. California's first inhabitants were maritime island peoples. Homecoming embraced the island, the coastal mainland, the channel, and the
village site. The traditional ecological knowledge of the *tomol* was
passed on across the generations for thousands of years.

Rich local knowledge of a region's animals and plants is fundamental
to human society. Local knowledge can generate respect for the natural
community as a higher language, a respect for the underlying unity of the
diversity of life in a region (Berkes 2012). Insight about your place can
translate into a renewed sense of inhabitation and community. Living-in-
place requires that we know where water comes from, where food comes
from, and where waste goes. A sense of place also requires that we be
aware of the circle of animals and plants that are native to a bioregion.
There are cultural representations of a place, such as the locations of
historical landmarks, the stories of animals and plants, the sites of Chumash ritual and ceremony, and the character of urban and rural areas.

Local knowledge is the collection of memories of those who inhabit a
place. The subtle change of seasons is reflected in the light-blue sky of the

fall or in the smell of California lilac in the spring. The colors and smells of each day are part of the ecology of California's Mediterranean heritage. Ecological awareness and local knowledge have long been described as key factors that contribute to the protection of special places (Orr 1992). Knowledge of a bioregion, for instance, can be determined by the use of the natural sciences (climatology, physiography, animal and plant geography), natural history, local knowledge, and the senses. "The final boundaries of a bioregion are best described by the people who have lived within it," according to activist Peter Berg and biologist Ray Dasmann (1978), "through human cognition of the realities of living-in-place." Understanding the needs of extraordinary animals like the wild southern steelhead or the brown pelican can also be a form of local knowledge.

THE AMPHITHEATER TO THE SEA

Our dependence on a healthy ocean extends across the blue horizon and connects diverse peoples and places across the sea. The Pacific Ocean is the largest ecosystem of the planet and covers 34 percent of the globe. The marine area hosts a very high diversity of seabirds and marine mammals. Many of the migrating species found off the California coast depend on a much broader ecological area that supports important feeding grounds in the northern Pacific Ocean. For example, the sooty shearwater nests in the southern islands of New Zealand but depends on the north Pacific marine area off southern California.

There are many coastal and marine areas of California that all together should be considered a Noah's Ark of the Pacific. For example, at least twenty-seven species of whales and dolphins have been sighted in the Santa Barbara Channel; about eighteen species are seen regularly and are considered residents (Dailey, Reish, and Anderson 1993). Over sixty species of marine birds may be using the marine and coastal waters of the region to varying degrees as nesting and feeding habitat, for wintering, and as migratory or staging areas. San Miguel Island supports the most numerous and diverse avifauna, with nine species having established colonies. Santa Barbara Island has several nationally and internationally significant seabird nesting areas, including the largest nesting Xantus's murrelet colony and the only nesting site in the United States of black storm-petrels. Brown pelicans maintain their only nesting area in California on Anacapa Island. Other seabirds, such as the sooty shearwater, visit the marine area to feed before returning to New Zealand to nest.

The presence of blue whales and other marine life depends on complex ecological relationships between the currents, the winds, the climate, and habitats. These relationships change over time and are often driven by subtle variation in sea surface temperature, as described earlier. The abundance and distribution of kelp forests is influenced by water temperature, as well as currents, pollution from the coastal mainland, storm activity, and the presence of other life forms, like urchins and their predator, the southern sea otter.

Scientists describe the synergistic and cumulative impacts of human activities on marine systems in the ocean off California. Human impacts, such as habitat loss and pollution, and climate change can influence the native species diversity of the region. Studies by Ben Halpern et al. (2009) at the University of California, Santa Barbara, include an analysis of the impacts of human use and climate-related factors on the California Current, which is the major oceanographic current off Santa Barbara. They show many areas of high impact and few marine areas of low impact. In particular, scientists have shown that parts of the near-shore marine environment include areas of high human impact. One reason is that near-shore marine areas are close to coastal watersheds, and pollution from rivers and creeks influences marine life. During heavy rain events, sediments and pollutants from coastal creeks and rivers can reach the northern Channel Islands. Studies of terrestrial inputs show that sediment plumes can reach as far west as San Miguel Island. These studies show the often subtle connection between the terrestrial systems of the region, such as a creek, land-use activities, and offshore marine areas. So we need to recognize that our treatment of the land and watershed influences the marine system.

For millennia, maritime stories have emphasized the need to return to the ocean as a source of wildness and nourishment for the human soul. Poets turn to the ocean not just for sources of food or aesthetic inspiration but also for what Henry David Thoreau referred to as "the tonic of wildness." This tonic of the wild sea is a foundation for maritime story and mythology. In one ancient maritime mythology, for instance, an albatross was referred to as a representation of the soul of a lost sailor at sea. An albatross is a central emblem in "The Rime of the Ancient Mariner," by Samuel Taylor Coleridge. In Coleridge's poem the bird is a metaphor for a burden or obstacle (as reflected in the phrase "an albatross around one's neck"), and in the poem a mariner is punished for killing an albatross. Throughout history, sailors who caught these birds let them free for fear of reprisal from the sea.

THE BIOREGIONAL MOSAIC

How we relate to the landscapes we inhabit is an essential part of home-coming. The five regions of the world with Mediterranean-type ecosystems are characterized by mild, rainy winters and hot, dry summers, are extraordinarily rich in biodiversity, and only cover 2.25 percent of the earth's land surface. These unique bioregions contain 20 percent of the named vascular plant species, which contribute to the life-giving values of healthy ecosystems and habitats (Stein et al. 2001). These regions are found in parts of Australia, Chile, and South Africa; in the California Floristic Province; and of course in and around the Mediterranean Basin.

The ecology of California's bioregions is influenced by natural and human-made features, and combines open space, rural, urban, and industrial features. In the early 1950s, sociologist William Whyte (1958) described the urbanization of Southern California as "urban sprawl." Urban sprawl and suburbanization continue to threaten ecosystems and increase pressures on human communities. For instance, urbanization can fragment the landscape, which is known to impact biodiversity in many ways. In the Los Angeles Basin, there is very little wild nature left—or even public space, for that matter. By 1959, only 3 percent of coastal Los Angeles could be considered wild. By 1995, only 1 percent of L.A.'s coastal habitats remained wild, while 84 percent of the landscape was urbanized (Davis 1990).

Most regions are a mosaic that includes a circle of native plants and animals within an urban, industrial, agricultural, suburban, and wild landscape—which together define and shape the ecology of a bioregion. This region is considered by scientists as a coastal province that includes a number of rich *ecotones*, where diverse ecosystems intermingle and overlap—the overlaps between island and marine ecosystems, marine and coastal ecosystems, and the coastal valleys, foothills, and mountainous landscapes represent transition areas that are home to the highest native-species diversity in the United States. The diverse geology and topography are important factors in the rich biodiversity of the region.

Identifying with the animals, plants, and creeks of a region is a good place to begin to develop a sense of place. The region combines the realm of human culture and the realm of climate and ecosystem—as shared habitat of humans, plants, and animals. Making the connection to a region can involve an extension of human identity: it starts with the identification of the neighborhood creek. In time, this identification with the creek can be extended to the animals and plants of an entire watershed. Poet and philosopher Gary Snyder (1995a, 1995b) refers to "watershed consciousness" as a way of thinking ecologically about the creek, river, and channel. The presence of healthy forests is supported by the creek's flow; the river's flow and habitat are home to songbirds; the blue heron is feeding in a wetland, but flies upriver to hunt along the mudflats and riparian areas; the river carries sand and sediment, which empties into a beach area; the beach area is used by families; the river's sand and sediment are deposited in the marine area and influence the productivity of a kelp reef, which is fished by a commercial fisher. The river is connected to the forest habitat; the fish and birds are dependent on the quality of the habitat in a coastal river's bank; beachgoers depend on sand carried by a river or creek to shore; and the farmer depends on the quality of the soil replenished by the periodic flood of a river. It is these linkages and connections that are the most important to sustain.

The bioregional mosaic comprises interdependent parts: sea, creek, farm, mountain, forest, private residence, grassland, soil, industrial site, and foothill chaparral. Bioregions include a mosaic of urban, rural, and natural characteristics. Imagine that you are at the top of one of the mountain trails along the coast. From the rugged landscape of distinctive sandstone monoliths and dense chaparral you pass through shaded canyons and feel the coolness of riparian bay laurel and sycamore. If you were to follow a creek down to the sea, you would probably enter a farmer's avocado, orange, or lemon orchard. Later, you would likely enter a neighborhood with backyards backed up to the creek, and then other urban and industrial developments. Some parts of the creek would be paved and the riparian vegetation removed, others still wild, with pools supporting fish and frog. Still following the riparian edge of the creek you may eventually reach a wetland or estuary, where you might be lucky to see a blue heron hunting along the banks of the slough or shorebirds working the edges during low tide as the waters drain to the sea.

In many watersheds across California, I have followed the path of the salmon from the coast to the upper foothills and mountains of the coastal range. My particular focus has been on the plight of wild salmon. The steelhead is a migratory salmon that depends on healthy coastal watersheds. They are born in freshwater streams; spend a portion of their lives in the ocean; and return to freshwater to spawn. In the early 1900s steelhead were abundant in the coastal streams in Southern California. Over the past century, human modification of riverine habitat has devastated steelhead populations in the region. In 1997, the National Marine Fisheries Service listed the unique southern steelhead as a federally endangered species. The service estimates the southern steelhead population to be less than 1 percent of its historic population size. The southern steelhead population has experienced the most dramatic decline throughout California and likely North America. The loss of freshwater habitat due to the construction of migration barriers such as road crossings, dams, and flood control structures in creeks and rivers is the single greatest limiting factor for steelhead in southern Santa Barbara County streams.

One essential part of coastal watersheds and aquatic ecosystems is the riparian habitat areas along a creek's or river's bank. Riparian habitats are critical to a range of species. Indeed, two-thirds of the federally and state-listed threatened and endangered species depend on aquatic habitats, such as wetlands, creeks, and riparian areas, during some part of their life cycle. Of all the United States, California ranks second in the number of aquatic species that are listed as threatened or endangered.

Birds, mammals, and a range of other species rely on healthy watersheds and associated wetlands for food, freshwater, and shelter, especially during migration and breeding. It is important that we recognize the complex and dynamic relationships between coastal and marine ecosystems of the bioregion that are essential to sustaining keystone and indicator species, such as birds and mammals. Wetlands are physically linked to a watershed by the delivery of water, sediment, and nutrients to the wetland from the watershed. The importance of this delivery cannot be overstated. Within a particular geologic context, water, sediment, and nutrients from the watershed define the type of coastal wetland that emerges or the quality of the agricultural soils in the river valleys.

Sacred Places

If you inhabit a place, you can gain a greater understanding of the sacred features of the landscape and seascape. I live in the midst of old oak trees. The branches of these old souls cover the house. After a cold front passes, the winds blow from the sea to the mountains. The oaks respond to the wind by dropping acorns. I think the mountains also respond to these winds in all the colors of spring. The oak grove is a whole interdependent system of relationships—it is the relationship between the trees, the entire grove, the canopy and undergrowth provided by the habitat, the birds (and their songs), and all the others that is sacred.

To protect oaks is a sacred and ancient activity. The Roman poet Ovid said, "Here stands a silent grove black with the shade of oaks; at the sight of it, anyone could say, 'There is a god in here!'" Sacred groves are important to many peoples, in India, Sub-Saharan Africa, parts of the former Soviet Union, East Asia, and Oceania. In many places, oak groves are considered to belong to the gods. In the Western Ghats, a mountainous region along the seacoast of India, a sacred grove (called Devaravattikan) exists in the midst of an agricultural zone. Devaravattikan remains a temple to the people. It is believed that the gods live among the trees, so people enter the grove with respect. This sacred grove helps the people maintain an ecological relationship with wild living creatures. *Kans*, the sacred groves of India, are a refuge for native diversity and remain the only habitat for a number of species of small animals.

Sacred groves were also once important features of villages in Europe, Mediterranean landscapes, and pre-Columbian America. These groves were sacred parts of a living and native culture.

The presence of oaks is one key factor contributing to the richness and diversity of life. The smell of oak pollen is in the air. It remains a sweet smell that I will eternally embrace. The oaks represent the flesh of the valley—the body and torso of my watershed.

TURN TO THE MOUNTAIN, THINK OF THE SEA

Early one evening I was driving to my home in the Santa Ynez Mountains, into the range that sets the stage for the region. I saw a large bobcat lying in the middle of the street, and pulled over to the side of the road. The bobcat was still panting and breathing. I cautiously petted its thick fur coat. Its ears, teeth, paws, and eyes were large, for hunting small prey. I thought of the bobcat looming in the shadows of the brush, waiting for the appropriate time to cross the road.

The young bobcat died. I took it into the hills of its origin to bury it under an old oak tree, near coyote brush, monkey flower, and coastal sage. The bobcat is now part of the soil and the oak tree. The soil, bobcat, and oak tree are linked. They are part of the breath of this region and landscape. The bobcat reminds us that we are not far removed from the wildness of this region, place, and community. The soil is made up of the flesh and bones of every creature that shares this place with us.

The animals, plants, and soils are the gifts of a region. The life-giving values that are carried by healthy ecosystems depend on the flourishing of wilderness. The sound of a wild sanctuary of shorebirds feeding in a coastal estuary embodies spiritual and sacred significance. The challenge is to translate these intrinsic values associated with natural places into our relationships with a place and community. This is not necessarily the job of government agencies or elected officials. A large part of the preservation of wild lands will require new community-based partnerships that cut across the boundaries of landownership, involving public lands and privately held areas. Collaboration with farmers, ranchers, and resource users is an essential prerequisite for stewardship.

Wilderness is not a distinct area beyond the horizon, or bounded by a fence, or separate from the more urban or suburban parts of the landscape. In this sense, wilderness is not detached from human activities, suburban neighborhoods, or the grove of lemons or avocadoes. Many songbirds depend not just on the coastal sage scrub, native grasslands or chaparral ecosystems but also on suburban landscaping. Important predators, such as bobcats, coyotes, and mountain lions, depend on connections and linkages that cut across the landscape mosaic. Accordingly, we need to understand that human activities, such as a road or culvert in a creek or the use of pesticides and other pollutants, can threaten the habitats and relationships that need to exist for animals.

In recognizing the importance of wild places, one essential question remains to be answered. Can we pass from institutions in support of multiple uses to a form of governance that is sensitive to, protects, and conserves the multiple values that are carried by ecosystems, integrating human culture with these values? The strength of any truly adaptive culture is based on the value orientation(s) of decision-makers; the scale of resource use; the level of native species protection that is supported and maintained over time; the ability of decision-makers to learn and respond to new information and values; and the communicative and integrative skills of the practitioner, the resource user, the scientist, and the citizen. The maintenance of ecosystem services—such as the presence of pollinators, healthy soils, and clean water and air—is derived from the interaction between natural and the built environments.

For the past several years, I have been visited by an island warbler (from Santa Cruz Island) who has flown across the channel to sit outside the window of my house in the foothills. The island warbler has been looking in on me, so I decided it was time to visit the home of the warbler. From a small beach on Santa Cruz Island, I am looking out across the Santa Barbara Channel. I can see the Santa Ynez Mountains, and Pine Mountain behind the range. It is a unique vantage point that reflects the unique connection that we share with the northern Channel Islands, and membership in a community that includes the animals, plants, and insects of these islands. Unique habitats like bishop pine forests are slowly recovering from the impacts of nonnative species. The feral goats and pigs have been removed. These forests depend on the occasional fog that provides the moisture needed for growth and renewal. The rewilding island reflects the regenerative power of species to overcome and adapt.

From the island, one can gain a deeper perspective of the wonders of the intimate relationships between human beings and the natural world. Without this intimate connection to a place, we become more vulnerable to the changes we will likely face in an age of climate change. As with the coastal mainland, the Channel Islands are vulnerable to change from afar—the changes from climate-related and natural forces, such as subtle changes in the currents and eddies of the marine environment, are beyond our capacity to control. As in the past, we need to be prepared to adapt lifestyles that can sustain these unique facets of our respective regions. This requires that we become more attentive to the landscapes we inhabit, more aware of the system we depend on.

We need to be more attentive to ecosystem functions, such as nutrient regulation in the soil and water, that are provided by the traits of organisms. We receive many benefits from functioning ecosystems. A good example of the importance of one ecosystem service is the pollination provided by species such as bees and birds. Without interaction between animals and flowering plants, the seeds and fruits that make up nearly 80 percent of the human diet would not exist. In *The Forgotten Pollinators* (1996), Stephen Buchmann, one of the world's leading authorities on bees and pollination, and Gary Paul Nabhan, an award-winning writer and renowned crop ecologist, describe the vital but little-appreciated relationship between plants and the animals they depend on for reproduction—bees, beetles, butterflies, hummingbirds, moths, bats, and other species provide these essential services that we often take for granted. In many instances we actively exterminate them as so-called pests.

THE PROCESS OF RE-INHABITATION

Love is where attentiveness to nature starts, and responsibility
towards one's home landscape is where it leads.
—John Elder (1998, 13)

Engaging in civic life in a community requires a values-driven approach, designed to advance principles of responsibility and stewardship that are locally grounded in the social, economic, and ecological components of a region. It can combine stewardship perspectives that can sustain rural and natural lands; support the integration of skills and actions that can contribute to the future restoration of unique features of the landscape; and rebuild human communities and reconnect them to the natural systems they are dependent on. Place-based civic engagement can support a stronger linkage and relationship between the social and economic aspects of a region that are often disconnected from nature, reconnecting ordinary citizens to the responsibilities of community stewardship and civic life. Civic engagement promotes civic knowledge, responsibility, and participation in individual and collective actions in support of strong networks and social alliances that can sustain a community's biocultural values.

The process of re-inhabitation requires a renewed sensibility that is supportive of Aldo Leopold's ethical maxim: "A thing is right," he writes in *A Sand County Almanac* (2001), "when it tends to preserve the integrity, stability, and beauty of the biotic community. It is wrong when it tends otherwise." As Leopold describes further, "The land ethic simply enlarges the boundaries of the community to include soils, waters, plants, and animals, or collectively: the land." The "land ethic" recognizes that individual species have moral standing; but a greater, more unified and coherent practical principle is one that recognizes the importance of maintaining the integrity of ecosystems as specific places and as essential to the sustainability of the more-than-human community. The "land ethic" is a more holistic approach to the treatment and preservation of biodiversity insofar as it supports the moral consideration of entire ecosystem processes that human beings depend on for essential services.

Leopold is considered one of the fathers of the restoration movement. His land ethic was based on his personal journey as a practitioner and scientist in environmental management in an era that emphasized the use and development of natural resources. His worldview was influenced by direct participation and knowledge of the landscape. His work includes stories about the wonder of the natural world, and what is needed to protect it. Change in one's value orientation, for Leopold, was the key to protecting species. His concept of the biotic community is one foundation for the contemporary conservation movement—it includes a value orientation that extends to the great circle of animals, plants, and insects that are part of the landscape. The story of the special places also needs to be told. Remembering the story of wild steelhead in creeks and rivers, for example, ultimately depends on a shared capacity to care for those coastal areas they depend on for their survival.

CHAPTER 3

Re-inhabitation

Watershed-Based Activism in Alta California

The watershed is the first and last nation, whose boundaries,
though subtly shifting, are unarguable. Races of birds,
subspecies of trees, and types of hats or rain gear go by the
watershed. The watershed gives us a home, and a place to go
upstream, downstream, or across in.

—Gary Snyder (1995a, 65).

The magnificence of a wild salmon's run up a river is hard to forget.
Most species of wild salmon are anadromous, which is derived from the
Greek *anadromos*, "running upward." An anadromous species is born
in freshwater and after several years in the marine system migrates
upriver to spawn at the same location. This homing behavior depends
on the salmon's olfactory memory, which refers to the recollection of
odors. The odors of a specific creek or river within a vast watershed are
deeply embedded in the genetic makeup of salmon. In this sense, a salmon's drive upstream from the sea is based on a sensory memory of the
particular watershed. The river is a genealogical lifeline that connects
freshwater habitat, spawning ground, marine system, and salmon. In
reproduction, this profound sense of place is passed on across generations of salmon.

We have much to learn from the return of wild salmon. In Northern
California, indigenous people refer to the king salmon as Lightning Following One Another, Chief Spring Salmon, Two Gills on Back, Quartz
Nose, or Three Jumps. Wild salmon represent a totem of threatened
watershed ecosystems. The term *totem* is derived from the Ojibwa word
ototeman, "one's brother-sister kin." The Ojibwe peoples are a major
component group of the Anishinaabe-speaking tribes, a branch of the

Algonquian language family. The reference to salmon by indigenous peoples also reflects their profound understanding of the ways of salmon and their understanding of a watershed.

Our relationship with the greater watershed influences the future return of salmon. In the Klamath-Siskiyou bioregion of Northern California and southern Oregon, the diverse indigenous peoples of the Yurok tribes created many years ago a salmon dance to galvanize the region's inhabitants, and to resolve their conflicts over the lack of salmon during periods of drought. Similar watershed-oriented ceremonies have long been traditional among peoples who depend on salmon as the primary source of food. It is believed that without the dance, the watershed's salmon will not return. Traditional knowledge holds that a river is a lifeline that connects people with place. Making the commitment to responsibly care for the greater watershed can galvanize diverse human communities to protect salmon and a range of other aquatic species.

The salmon are resilient species that adapt to changes in the river's flow and the greater watershed ecosystem. Watersheds and the species that depend on them are influenced by periodic droughts and climate-related events, such as flooding and fire. In Southern California, steelhead salmon are the most threatened and endangered species of all West Coast salmon. South coast steelhead evolved to adapt to extended periods of drought. The rainbow trout is a steelhead that does not return to the sea, and remains in the upper watershed or in the deeper pools of water along a creek or river's corridor. In summer months, a river or creek in Southern California rarely reaches the sea. So south coast steelhead cannot make it up or down the river or creek. When the river's flow returns during the winter or early spring, the trout can spawn with the returning steelhead. Rainbow trout can also return to the sea. The wild steelhead is different from other species of salmon insofar as they can change their spawning habitat to a different creek or river area. As the ice retreated during the last ice age thousands of years ago, this particular south coast steelhead moved north along the West Coast to inhabit the rivers and creeks of Oregon and Washington. Every steelhead of the West Coast has a common genetic heritage that is based on the south coast steelhead species.

My fascination with the path of the wild salmon led to over a decade of research and writing on the politics of restoring wild salmon to the watersheds of the West Coast. During the late 1980s, hundreds of millions of dollars were spent on plans to recover and enhance wild salmon of the Columbia River Basin (McGinnis 1995). There are nineteen major federal dams and over a hundred smaller dams on the Columbia River, which have transformed the river's flow to serve the economies of the Pacific Northwest. The water behind the dams is used for recreation, flood control, irrigation of farmland, transportation of agricultural products like wheat or barley, and hydropower, and is a source of drinking water for cities and suburbs. This multiple use of the watershed has failed to consider the ecological values that are associated with a river system with respect to the needs of wild salmon. For instance, the salmon needs to traverse each dam that blocks the river's flow. Irrigation systems that remove freshwater for agriculture in the San Joaquin Valley of California carry dead smolt or juvenile salmon. Extinction is a possibility for most wild salmon. Our reliance on hatcheries also adds to the many threats to the survival of wild salmon. A hatchery-produced salmon is not a wild salmon, and masks the decline of a number of wild species. Hatchery salmon introduce diseases that can threaten wild salmon, and they also compete for resources that wild salmon depend on.

One of the most dramatic examples of the challenges salmon face in their migration up the Columbia River is the Redfish Lake sockeye salmon. Sockeye were listed as endangered in November 1991, and were the first Idaho salmon species to be listed by the National Marine Fisheries Service. They are unique among sockeye. They travel more than 900 miles and climb more than 6,500 feet in elevation, and they are the southernmost North American sockeye population. In the 1880s, observers reported lakes and streams in the Stanley Basin teeming with redfish. There was talk of building a cannery at Redfish Lake. Returns were estimated at between 25,000 and 35,000 sockeye. In 1992, only four adult sockeye made it to Redfish Lake. One of the last male sockeye to reach Redfish was caught, its genetic material was collected, and it was prepared, stuffed, and mounted on the wall in the governor of Idaho's office. This sockeye salmon was called Lonesome Larry. When then-governor Cecil Andrus put the stuffed fish on his office wall, Lonesome Larry became the symbol of the entire Snake and Columbia salmon enhancement program.

THE WATERSHED COMMONS

To say that you are part of a particular watershed system means that you inhabit a distinctive living community and place. A watershed is defined by government resource agencies in many ways. Typical is the language of the U.S. Environmental Protection Agency (Browner 1996): "Watersheds are those land areas bounded by ridgelines that catch rain and snow, and drain to specific marshes, streams, rivers, lakes, or the groundwater." A creek, river, slough or estuary, according to this definition, is not a watershed. A creek or wetland is *part* of a watershed. Webster's *New Collegiate Dictionary* offers a similar definition: a watershed is "the whole region or area contributing to the supply of a river or lake; a drainage area." These definitions characterize the watershed as hydrologic entities, which continue to exist even if stripped clean of biota, soil, or culture. But a watershed should also be understood as a particular cultural identification.

Scientists, government resource agencies, and activists increasingly recognize the importance of a watershed-based approach to planning and decision-making. The general view is that a watershed-based approach provides one of the best units for planning because: (1) watersheds are ecologically meaningful; (2) they are spatially defined; (3) they can be nested hierarchically (small watersheds are part of larger watersheds); and (4) the health of an entire watershed can generally be measured by the health of the aquatic system. Our human capacity to make connections between mountains, rivers, and the sea is an important factor that contributes to our collective capacity to sustain our respective places and communities.

We rarely understand these connections, nor are we willing to nurture and protect the socio-ecological relationships that can support watershed communities. During the past eighty years, the utilitarian view of the river was one that supported the engine of progress. This instrumental value of a river has led to the development of some 75,000 dams in the United States—the literal rearranging of the waters of the continent. Dam building also contributed to the loss of cultural regionality: towns were drowned by the dam's reservoir, and people abandoned their homes. In the buzz of hydroelectric power development the more-than-economic values of the natural world are silenced. Each dam resonates with the technological treatment of nature as a factory. Drowning Hetch Hetchy to provide power for San Francisco redirected the downhill energy (potential energy being converted in nature into kinetic energy) into paths available for urban use (electric energy). Dams reflect an uncritical social reliance on modern technology, and a form of *spatial apartheid*—each dam separates unique ecological places (riparian areas, watersheds) to support developments that are mechanical yet human (irrigation, grain transportation, hydropower, urban development).

The era of dam building is over; we may be entering a new era of dam removal. In September 2009, the U.S. secretary of the interior announced an agreement between the California and Oregon governors and Warren

Buffett's PacifiCorp to remove the four hydroelectric dams on the Klamath River. The plan aims to restore the cultural and ecological integrity of the region. Dam removal on the Wells River, in Groton, Vermont, will restore the natural river habitat to free-flowing conditions, improve water quality and sediment transport, restore the river channel, and increase and improve fish and wildlife access to habitat for resident and migratory species such as eastern brook trout and Atlantic salmon (known as Yankee salmon). Removal of this dam in 2014 opened a total of 22.31 miles of stream, 6.4 miles of which is cold-water habitat upstream from the dam. Dam removal began on the Elwha River in mid-September 2011. The Elwha Dam is gone in Washington; over 50 percent of Glines Canyon Dam has been removed; and the Lake Mills and Lake Aldwell reservoirs have drained. The Elwha River now flows freely from its headwaters in the Olympic Mountains to the Strait of Juan de Fuca for the first time in a hundred years. For over a decade, plans to tear down the Matilija Dam, north of Ojai in Ventura County (California), were moving forward. The $140 million project stalled a few years ago. Today, a strong coalition of watershed activists continue to work with government agencies to bring back southern steelhead to the Ventura River, and remove the more than 2 million cubic yards of fine sediment built up behind the dam.

No turnaround is more amazing than that of Idaho's Snake River sockeye. Since 2008, more than 650 sockeye have returned annually to the Sawtooth Valley, peaking in 2010 with 1,355, the most since the 1950s, before four dams were built in Washington. One reason for this hopeful return of the sockeye to Redfish Lake is found in the cultivation of a diverse watershed-based movement across many regions of the United States. Watershed activism is a particular form of place-based activism, combining the science of watershed dynamics with the place-based sensibility that is required to protect an entire ecosystem of relationships. A watershed is a culturally meaningful construct because of the associations, relationships, and partnerships that can be created. This chapter explores the role of science and values in this diverse watershed movement. If taken seriously, watershed-based thinking, organizing, and planning can radically change the way we confront socio-ecological issues. Grass-roots, community-oriented, and watershed-based organizations have proliferated in California. Government and non-government watershed organizations and councils include among their members community activists, private property owners, representatives from local, state, and federal resource agencies, industrial interests, restorationists, and conservationists.

In the context of climate change, the future of watershed ecosystems and their salmon remains uncertain. Californians are facing one of the most severe droughts on record. Governor Brown declared a drought state of emergency in January 2014, and directed state officials to take all necessary actions to prepare for water shortages. The drought will surely affect the future of wild salmon and other native species in California's rivers and creeks. It will also reshape the agricultural industry and challenge cities to develop alternative sources of water. California collects water in more than 1,200 large reservoirs and moves it around through the world's largest aqueduct machine. The complex network of waterworks and irrigation systems of the state contribute about 40 percent of the state's total greenhouse gas emissions. It takes a lot of energy to move water to the cities and agricultural areas of California. Currently, there are fifteen proposals across the state to build desalination plants to provide water. These desalination plants will also require major sources of energy to produce freshwater from the sea. One additional product of desalination is brine, which will likely be deposited in the near-shore marine environment. A brine plume in the ocean is not without its ecological impacts and can deprive the ocean of oxygen. One way or another, current and future water use in the state is irrevocably connected to the burning of fossil fuels, and to water pollution, and influences the general health and integrity of watersheds.

Today, salmon are increasingly threatened by the impacts of climate change. In particular, the long-term drought in California is impairing the ability of wild salmon to adapt and survive. From 2012 to 2014, California experienced the most severe drought conditions in its last century. The current drought in the state is the most severe in the last 1,200 years (Griffin and Anchukaitis 2014). Higher temperatures brought on by climate change and drought conditions will significantly harm California's salmon and steelhead populations (Kibel 2014). Ultimately, the recovery of wild salmon runs will require that we respond to climate disturbance at the watershed level, and ensure that there is enough water for salmon and other aquatic species to adapt to long-term climate disturbance.

EARLY DEVELOPMENT OF WATERSHED-BASED ACTIVISM

The concept of watershed-based planning is not new. In 1878, John Wesley Powell, the first head of the U.S. Geological Survey, proposed to

Congress that the political jurisdictions of new states in the semi-arid West should be based on watershed boundaries. For Powell, the watershed was the ideal management unit for a new form of governance, and he based his recommendation to Congress on the lessons he learned while travelling the Colorado River and meeting with indigenous people along the journey. He learned that the greater watershed's social and cultural identities were often based on the soft boundaries of a watershed. Traditional language and knowledge were based on the watershed system. But Powell's recommendations were not supported by Congress. The Colorado River was dammed. The traditional peoples of the Colorado River basin were displaced. Now the language and traditional knowledge of the greater Colorado watershed remain deep beneath the lakes and reservoirs; the river has been destroyed by the staircase of dams that block and control its once natural flow.

Powell's vision for a watershed-based form of governance represented the first wave of support for watershed-based planning in the United States. Another vision for a more ecologically grounded approach to planning is found in the work of Lewis Mumford. Mumford's classic essay "The Natural History of Urbanization" (1956) describes the dependence of cities on their hinterlands. The historical development of the urban landscape, according to Mumford, was one where the natural, rural, and urban features were separated by officialdom. For Mumford, an illusion of complete independence from nature and the agricultural areas on which the city historically depended is fostered by modern city planners.

Mumford's major contribution to early planning theory, *The City in History: Its Origins, Its Transformations, and Its Prospects* (1961), received the National Book Award. Mumford describes the growth of the mega-machinery of the metropolis—a system of planning and governance that separates the natural, urban, and rural features of a region, and contributes to the eventual decline of the city, its surrounding hinterland, and nature (Luccarelli 1995). Mumford showed the emergence of an "urban-rural perceptual divide," where distinct planning systems have emerged to support the development of the modern city—one for cities and one for rural lands. These systems of planning have different agencies, procedures, and remits for rural and urban spaces (and natural landscapes) through their division and zoning. Different land-use strategies, planning tools, and policy instruments are used by officials, with very little integration across natural, rural, and urban areas at the regional level: rural and urban landscapes are planned as opposites.

Rural areas in many regions are viewed by city and county planners as a convenient way to hold land until the time for urban growth.

The growth of the modern industrial city culminates in what Mumford (1961, 7) calls "the City of the Dead." But Mumford was also a utopian thinker and a visionary; he remained hopeful that a more sustainable urban–rural interface could be developed by adopting an ecologically grounded regional approach to land-use planning that embraced the linkage and connection between these landscapes. As the founder of the Regional Planning Association of America, Mumford advocated that urban and rural landscapes be planned as integrated parts of ecological regions or bioregions that could support a more holistic relationship between the natural, rural, and urban features of the landscape (Luccarelli 1995).

Mumford was one of the first public intellectuals to support a bioregional approach to planning. The regional planning movement of the early twentieth century was a loose connection of theorists and practitioners, such as Patrick Geddes, Benton MacKaye, and Mumford, devoted to developing an alternative approach to the design of future cities. In the early part of the twentieth century, the planning firm Olmsted-Bartholomew created a design for the future of Los Angeles that would have established parks, open spaces, and urban communities linked with gardens, and maintained the natural floodplain and riparian areas of the Los Angeles River. The plan was entitled "Eden by Design: The 1930 Olmsted-Bartholomew Plan for the Los Angeles Region" and was commissioned in 1927 by the Los Angeles Chamber of Commerce. After a major flood of the region, the plan was abandoned, and the Army Corps of Engineers implemented an alternative plan for the river that included flood control and thousands of tons of pavement (Davis 1995). The *once* river now resembles one of the city's freeways.

From the early 1930s to the late 1990s, a second wave of watershed-based planning included proposals, programs, and plans that were developed at local, state, regional (interstate), and federal government levels. This second wave focused on aquatic ecosystems, such as the development of formal interstate compacts for major river systems (McGinnis 1995). River basin authorities were established across the country to plan for the use of rivers. The literature and technical documentation during this second wave provided a set of generic principles to guide policy and program development. In addition, the scientific and intellectual basis of the watershed approach to policymaking was further supported by programmatic development by federal and state resource agencies.

During the 1990s, a third wave of enthusiasm and support for watershed-based planning emerged. An ecosystem-based approach to watershed planning was embraced by the Executive Office of President Clinton. There are four prevailing themes in the literature that supports watershed-based planning: the notion of boundary redefinition to deal with the spatial and temporal scale of ecosystem dynamics; issues and concerns related to scientific uncertainty and conservation planning; intergovernmental administration and coordination; and the development of principles of ecosystem-based management. In July 1998, the Western Water Policy Review Advisory Commission, created by Congress in 1992 to review federal activities associated with water resources, released a landmark report, *Water in the West: The Challenge for the Next Century* (Western Water Policy Review Advisors Commission 1998). This report was perhaps the most farsighted federal study of western water since Powell's proposal. And, like Powell's earlier proposal, the commission proposed a new government structure that reflects the hydrologic, social, collaborative, legal, and political "realities" of a watershed.

THE WATERSHED MOVEMENT

Throughout California the watershed movement is widespread and diverse. The Information Center for the Environment at the University of California, Davis, has an inventory of 660 watershed organizations. Most of these organizations do not address the entire scale of watershed dynamics but focus on particular areas within a watershed, such as a riparian area or wetland. This movement includes private property owners seeking to fend off greater regulation from higher levels of government, along with committed conservationists and restorationists seeking to protect aquatic species and their habitats. Watershed activists share a core idea: they recognize that they inhabit a distinctive place with a specific set of ecological connections. They have come, to some degree, to think and act in accordance with the boundaries of a watershed. There are watershed-oriented groups addressing the habitat needs of birds, fishes, and other aquatic species in wetlands, estuaries, riparian areas, and entire river basins.

There is both unity and variety in the watershed movement. Activists are engaged in efforts to protect native species diversity, and to promote preservation and restoration of habitats that have been degraded by human activities. One example in an urban setting is the Friends of the

Los Angeles River (FoLAR), founded in 1986 by Lewis MacAdams. Membership is open to anyone, irrespective of landownership, profession, or place of residence. FoLAR's goal is to protect and restore the last remaining natural portions of the river and to increase public awareness of the river. FoLAR is one of the oldest citizen-based organizations in Southern California devoted to restoration.

In the rural and natural setting of the Kings Range, the Mattole Restoration Council (MRC) was one of the first watershed organizations in the state. For almost forty years, the inhabitants of the Mattole Valley and the "Lost Coast" in California have worked together to restore watershed ecology and community. During the late 1960s, the Diggers and antiwar activists of Berkeley left the Bay Area to settle in the Mattole Valley. They considered themselves "second growth"—a metaphor for the second growth of the Douglas fir after the clear-cut of well over 90 percent of the Kings Range's forests during World War II. The homesteaders often built their homes on abandoned lots using recycled materials and the foundations of barns. As countercultural activists, they shared the philosophy of the "back to the land" movement of the period. Many of these new inhabitants of the Mattole would be the early writers and thinkers of the contemporary bioregional movement.

The workings of the MRC are worth exploring in more detail because this group is often considered the model of a watershed-based organization. In the Mattole, watershed activism is a means of incorporating the landscape and river into the human experience and the shared stories of place. During the 1980s, money raised by growing marijuana in the valley was used to support a network of on-the-ground restoration activities for the Mattole River's wild salmon. Without government resources of any kind, the early work of the MRC preserved the remnant populations of salmon that officials in Sacramento thought were extinct. Counting salmon in the tributaries and the creation of a low-tech hatchery for salmon were essential to the early success of the MRC in protecting three species of salmon. Over time, a number of citizen groups and organizations working to remove roads and restore forests, and devoted to public outreach and education, formed under the MRC. Its members combine knowledge of the river with activism, and social partnership with ecological awareness of the coastal range. In the 1990s, an educational curriculum for children of the region was established and focused on salmon ecology and recovery. Children carried buckets of juvenile salmon downstream. Murals in the school's playground showed salmon and other species along the river. Related activities included the theatrical group Human Nature, whose play *Queen Salmon* toured the Pacific Northwest.

The watershed ideal, in the case of the MRC, became a source of social identity for the community that inhabited the greater watershed area. The MRC shows that watershed-based activism is much more than just a sense of attachment to place. The watershed is part of what we are, and it can become what we wish to protect and defend. Watershed activism and organization can be a way of becoming "placed." Intermediate between focused restoration groups like FoLAR and the intensive, "bioregional lifestyle" of the Mattole inhabitants are a range of organizations linked in social networks. One key feature of those that were successful in protecting their place in the watershed was the establishment of social networks that combined many diverse organizations with the promise of developing a shared vision to support the watershed commons. For example, the Redwood Community Action Agency of the Watershed Information Network (WIN) in Humboldt County is a social network of diverse watershed-oriented groups and activities. WIN does not advocate a particular position. Rather, WIN's goal is to unify the voices of the bioregion by providing a safe place for diversity. WIN focuses on vocational issues such as training, and on building collaborative decision-making, providing expertise, scientific information, and a neutral public forum for negotiation and collaboration. WIN formed around the value of the process of restoration; this remains the network's common purpose and shared vision.

GEOGRAPHICAL ROOTEDNESS

Mapping place is an essential first step to self-governance and community-based empowerment at the watershed level (Aberley 1993). Mapping activities include the intuitive knowledge gained while inhabiting place and the sharing of information in community forums. In the case of watershed activism, maps are important forms of communication and relationship building, and they can foster trust and new partnerships that cut across political, economic, and administrative boundaries. Maps of a watershed often reflect a particular geographical rootedness and sense of inhabitation. Mapping necessarily involves the art, imagination, technology, and science of defining the region. This art, craft, and science of mapping the diverse layers of a watershed are described by David Robertson (1996, 1015): "As one moves from geography to biology to culture, subjectivity increases. Since the geographical map is the most straightforward, it is best to begin with it. The trend at present is to set bioregional boundaries by watershed . . . or landmass. . . . The biological map is an overlay with information about vegetation types and animal ranges. The cultural map is an additional overlay that takes into account political, economic, and social borders."

Maps should be the product of community engagement processes that foster collaboration across diverse interests and values. A map of a place is as much a cognitive as it is a technical exercise. A community-based mapping process involves an exchange of energy, materials, information for planning, and other societal elements. In a characterization of the MRC, Freeman House (1999) describes the evolution of the diverse members of the council as a means to extend social identification with the ecology of a place. At first, the MRC focused on restoring the lower parts of the habitats along the Mattole River. In time, the members recognized that the scale of their activities did not go deep enough into the headwaters of the Mattole. The scale of their activities expanded to include responses to the lack of healthy forests in the headwaters, planting native species, and decommissioning roads that were contributing to high amounts of sediment in the river. Expanding the scale of watershed-based activities required new partnerships and alliances. In *Totem Salmon: Lessons from Another Species,* House explains this learning process of extending identity well beyond a family, locale, and the neighborhood. "As we engage directly the recovery of our shared habitats, we find ourselves in the embrace of the expansive community that offers the best hope of realizing ourselves as fully human. There is no separate life" (218). A deeper awareness and sensibility of the needs of an entire watershed emerges over time. Mapping the greater Mattole watershed was a crucial part of this process.

VALUES AND SCIENCE IN WATERSHED ACTIVISM

The watershed movement embraces a diverse set of organizations and interests, including traditional conservation groups, restorationists, Earth Firsters, local development interests, salmon advocates, private property rights groups, scientists, and bureaucrats at all levels of government. The watershed movement is pluralistic and diverse. There are a variety of motives and goals that propel participation and membership in watershed organizations.

With funding from three awards from the National Science Foundation, my colleague, John Woolley, and I conducted a number of studies of watershed activists and organizations (McGinnis, House, and Jordan 1999; McGinnis, Woolley, and Gamman 1999; Woolley and McGinnis 1999, 2000, 2002). In one of our studies on the California watershed movement, we used a number of different methodologies, including case studies and surveys of members of watershed organizations. To gain a better understanding of the general characteristics of California watershed activism, we explored the motivating forces and values, the role of scientists and scientific information, institutional factors, and the practice of river restoration.

One focus of our research is worth describing in more detail. We conducted several surveys of watershed and river restorationists with respect to their environmental worldview. We used statements to assess environmental attitudes on a scale formulated by Riley Dunlap and associates (2000) after several decades of research on the topic. Dunlap and others are considered environmental sociologists, and show that there is an ecological worldview that reflects an ethical view that is nature-centered, as opposed to human-centered, and supports a value of nature as more than a commodity to be traded and used.

To begin to understand the role of environmental attitudes in watershed activism, we began the process of identifying watershed organizations using the online Surf Your Watershed service created by the U.S. Environmental Protection Agency (EPA). From 1998 to 2002, the number of reported watershed groups in the state increased 300 percent. We consulted references, and then contacted the groups listed on the service by both mail and telephone. We supplemented group and contact information from two later lists of watershed groups: the California Watershed Project Inventory and a list prepared by For the Sake of Salmon. The former listing was quite extensive, with over 395 entries. We found that the loss of groups and turnover of representatives was high. But, as the watershed movement has gained prominence and visibility, so too has the number of organizations voluntarily identifying themselves as a watershed group.

There is an extreme degree of organizational fluidity in the movement. Watershed activism is dynamic, and involves substantial change in groups and in membership. Watershed organization includes many social alliances, networks, and cross-linkages. While this is probably true of almost every significant social or political movement, we must not lose sight of the fact that our research involves a snapshot of a changing "movement."

Our study sample included 463 individuals who, we believed, belonged to 193 different watershed groups. These groups, in turn,

were drawn from 70 California watersheds. We sent out mail question-
naires to watershed activists. The responses involved 217 individuals in
98 watershed groups from 45 different watersheds (for response rates of
0.47, 0.51, and 0.64, respectively). Fully 20 percent of our respondents
informed us that they were involved primarily in a watershed group different from the one through which we had identified them. The median
respondent claimed membership in two watershed groups. Our sample
is broadly representative of the relevant underlying populations. There
are 153 USGS watersheds or "cataloging units" in California, of which
124 (or 81 percent) have at least one self-identified watershed group
(according to the EPA's Adopt Your Watershed program).

At the time of our study most watershed groups were located in urban
settings with relatively high population density. Those areas without
watershed-based organizations appeared to have slightly poorer overall
water quality and fewer aquatic species at risk. There was a fairly strong
association between the number of endangered species in a watershed
and the extent of watershed group activity.

Watershed groups in California are, as a generalization, located
where the people are. The largest third of watersheds by population
included 57 percent of the watershed groups listed by the EPA, 60 percent of our sample, and 62 percent of our respondents. The largest half
of watersheds by population included 74 percent of all watershed
groups, 75 percent of our sample, and 69 percent of our respondents.
Geographically, the largest numbers of watershed groups were in
coastal Northern California, in major metropolitan areas, and in areas
known for their natural beauty. Relative to the population size, we
found that the highest rates of watershed organization were not in the
population centers of the coast but in the relatively rural watersheds of
Northern California and the Sierra.

Our study showed that watershed organizations were involved in a
number of activities, from cleaning up the river or creek to participating
in formal planning and political processes. One major finding of our
survey research was that the age of the watershed organization influenced the types of activities that watershed organizations were involved
in. A second finding from our survey of members of watershed organizations showed that the presence or absence of an endangered species contributed to a number of these activities. Restoring aspects of the watershed was often supported by the presence of an endangered species.

With respect to our exploration of the attitudes of watershed activists, the survey revealed broad areas of consensus among activists that

were substantially ecocentric. Respondents evaluated a battery of statements modeled after the new ecological paradigm developed by Dunlap and others. The responses from those surveyed in our study reveal ecocentric values among watershed activists. For example, the strongest mean response of agreement, 1.37, was for the statement "Nature has other than economic value." This is a very direct expression of ecocentric views. The strongest disagreement, 6.6 on the 7-point scale, was for the statement, "Natural resources should be used for the benefit of the present generation." This is a flipped version of the more familiar ecocentric expression—which also happens to have elicited the second-highest level of agreement in our survey—"Natural resources should be preserved for the benefit of future generations." The statements that elicit agreement are expressions of general values about our relationship to nature and our responsibility for it. They are not about the state of our knowledge or the means to action.

Of particular interest were the statements in our survey that touch on the values associated with science and scientists. Respondents disagreed among themselves about whether they know enough to manage watershed ecosystems responsibly. Forty-six percent agreed (or somewhat agreed) that they know enough; 44 percent disagreed, maintaining that they do not know enough. The irony of this is that respondents to the survey believed that good science supports their views. Of those who agreed that they know enough, 88 percent believed that good science supports their view. Of those who disagreed, 77 percent thought that good science supports their view.

We also found that a watershed is a social construct. Where, *really,* is "the watershed" located? This is a question that has epistemological and physical significance. A watershed is not merely defined in terms of the distinct hydrological features, but, as the MRC reveals, the watershed has social and epistemological significance as well. The inherent ambiguity of the definition of "a watershed" encourages misunderstanding about the topic and conflict over scope. One's sensibility to place, and a major river or creek, is an essential feature of the watershed movement. In the context of watershed activism, the watershed is as much a social construct as it is a hydrologic unit. The perceived adequacy of science varies tremendously from place to place. Within particular watersheds, activists equally committed to the idea of scientific knowledge disagree on the adequacy of current knowledge. Useful knowledge is found not in formal scientific sources or even from "scientists" but from local citizens. The primary source of information that members of watershed groups most often depend on is people with local knowledge; these are the members who are the most trusted.

In watershed-based planning it appears that science truly is "situated" within a particular local context and the vision arising from the experience of groups of activists. Our findings about science, use of information, and sources of activism in the watershed movement resonate with Donna Haraway's (1988) notion of "situated knowledge" and her argument for "epistemologies of location." There are three central claims within the concept of situated knowledge: knowledge and experience is situated; multiple visions and values are accepted within a particular context and situation; and there is a multiplicity of experiences of truth. Among watershed activists, we found evidence of such situated knowledge.

CULTIVATING A WATERSHED CONSCIOUSNESS

After over a decade of studying watershed organizations and their planning activities along the West Coast of the United States, I have learned one important lesson: there is an emerging ecological identity that watershed-based activists share. In time, the social identification with a place is one common feature of watershed activists. Like the salmon, we cannot avoid interacting with and being affected by our specific location, place, and bioregion. The watershed lends itself well to the development of community-based planning. There is a diverse watershed-based movement underway that represents a renaissance of community relationship building. Because a watershed is a place, it can help illuminate a wide array of issues and concerns associated with the more-than-human community. The watershed can link us in both ecological and social ways to the communities we inhabit. House (1999, 159) notes: "The living region requires of us that we become its intimate inhabitants, and further, that we regrow our sense of community as a function of that inhabitation."

There is the human need to develop an "initial" communal relationship with place—to work with one's community to re-inhabit a watershed commons. Working with other members of a community is a trait of humanity, even though it seems to have been lost with respect to our economic modes of consumption and production. Hands-on land and watershed restoration of the habitat along a creek or river or wetland can be one means to restore the connection between the mind, the body, and the flesh of the world we inhabit. Watershed-based activism can mean starting to work for a fuller and richer range of community-based opportunities and experiences that, in the future, can transcend the production-consumption economies of globalization.

Despite the pressures of globalization, we can resurrect a communal relationship with the watershed, drawn from the senses, memories, and stories we share in a place. In new social and ecological partnerships, a community can restore a sense of place that has a lasting force. This vision starts with the recognition that the self is not just an autonomous consumer and producer in a global economy but an inhabitant of a place and a participant in an earthly community. A good way to start thinking about your relationship to the greater watershed is to engage and interact with it. Participate in the river's flow, swim upstream or downstream as if you were a salmon, and cultivate a sensory and olfactory memory of place as if your livelihood (and that of your family and neighborhood) depended on it. The consciousness that emerges with direct human participation in place renders the science of ecology as a sensibility and ethic that can support the watershed. Respecting and caring for the watershed commons is an essential first step to address and respond to the many social, economic, and ecological threats and impacts that we have on ecosystems and the places we inhabit.

CHAPTER 4

A River between Two Worlds

Watersheds and Wastesheds of Aotearoa (New Zealand)

Nā Tāne I took, ka mawehe a Rangi rāua ko Papa
nāna I tauwehea ai, ka heuea te Pō, ka heuea te Ao.

It is by the strength of Tāne
that sky and earth were separated, and light was born.

—Traditional Māori teaching

In 2010, I accepted a position at Victoria University of Wellington and moved to New Zealand. My primary goal was to gain a greater understanding of the coastal and marine ecology of the Pacific Ocean. So I left my home in California and spent nearly three years working on a government-funded project to provide recommendations to the country's ministries on how to strengthen their ocean governance framework.

Aotearoa ("land of the long white cloud") is the indigenous Māori term for the islands of New Zealand. I learned that if you stay in a place long enough the once unfamiliar landscape can begin to speak to you; and the knowledge of the place can slowly change your understanding of who you are and where you are. In contrast to the dry river beds of Southern California, the major rivers of New Zealand flow year-round.

There are several prominent mountains on the North Island; Ruapehu, Ngauruhoe, and Taranaki are the highest. The first two are within Tongariro National Park, in the center of the North Island. Taranaki stands alone on the west coast, facing the Tasman Sea (figure 4.1). These mountains are part of Māori mythology, which portrays Taranaki as a warrior mountain in battle with Tongariro, Tauhara, and Pūtauaki. Also in the area was the maiden Mount Pihanga. According

FIGURE 4.1 Mount Taranaki, New Zealand. (Image: M. V. McGinnis.)

to the myth, a great battle between these mountains for Pihanga's love erupted in fire and smoke. The earth moved. After the battle, Mount Taranaki moved to the coast, and along the way he created the river valley of Whanganui and the wetlands of Te Ngaere. It is rare to see Taranaki not covered in a rainy fog. In the legend, the rain means that Taranaki is weeping over his lost love, Pihanga. When Taranaki is lit up by the sun, he is showing his strength, and reminding the other warrior mountains that he still stands.

One consequence of the "battles" between the mountains of this bioregion is that the soil of New Zealand is deep in volcanic ash—along with bird guano accumulated over thousands of years. As I walked toward the headwaters of the river I thought of the eruption of the region's Taupo Volcano. The explosion of Taupo is the world's largest known volcanic eruption of the past 70,000 years—an eruption with a volcanic explosivity index of 8. The explosive force of this eruption was greater than the combined power of all the nuclear weapons ever created by humanity. It occurred around 24,500 B.C., in the Late Pleistocene. Tephra from the eruption covered much of the island with ignimbrites up to 200 meters deep. Most of New Zealand was affected by ashfall, with an 18 cm ash layer left even on the Chatham Islands, 1,000 km away. One consequence of the eruption is that virtually every living plant and animal on the island was destroyed. Modern Lake Taupo partly fills the caldera created during this eruption. A 140 km² structural collapse area is concealed beneath the lake, while the lake outline reflects coeval peripheral and volcano-tectonic collapse. Later erosion and sedimentation had long-lasting effects on the landscape, and caused the Waikato River to shift from the Hauraki Plains to its current course through the Waikato to the Tasman Sea. The ash mixed with the thousands of years of accumulated bird dung and the seeds of native plants, which, in time, led to the landscapes settled by Māori when they arrived by canoe (or vaca) over 800 years ago.

Near the summit of Taranaki, the sun began to fall, and the fence posts lining each paddock for grazing cattle or sheep were less visible; the boundaries between the sea, water, and land were less clear. The lines between places are not as clear as we presuppose. I recognized the oceanic spaces that connect us, the wake-up calls of migrating birds and their songs, a meandering river's path, the storm's whisper of wind, and the saltwater in my hair.

Early that evening, I wandered down from the headwaters of Taranaki. Off trail I met a farmer. "I can hear, smell, and see the river," the dairy farmer told me, "and so can my cows." I could see the wild brush of the country that remained within a landscape mosaic of dominating green dairy pastures, and the riprap of a transformed Patea River, with eroding banks and no vegetation. The native brush had been cleared, and in its place grew European grasses that were fed on by dairy cattle. Dairy plots fragmented the landscape and cut across the meandering river. The farmer's large dairy herd extended to the horizon. Cattle were in the river, crossing the roads, eating the remnant native brush. I could not hear the songs of birds. I wondered how resilient the river and its greater watershed are now that industrial agriculture has taken hold. After processing, the milk would be shipped overseas to China. Nonnative European grasses fed by fertilizer and groundwater lay beneath my feet. I walked two diverse and conflicting landscapes of the river, two "kingdoms of force" (Marx 2000) and two terrains of consciousness, a mosaic of human-made and natural surroundings. I could sense the volatility of the changing place.

When Captain Cook arrived in New Zealand, his crew returned to the ship because they could not sleep: the bird chorus was too loud across the landscape. The richness of the soil caused by this mixing of guano and ash is the foundation for the grassy richness of the landscape that feeds New Zealand's dairy cattle. The richness of the soil and the bounty of the marine systems of New Zealand led to early trade and exploitation of resources by European settlers. Colonization of the islands across Oceania (the region of islands that include Melanesia, Micronesia, and Polynesia, along with New Zealand) was supported by a myth of an ocean and its rivers as an expanding frontier with endless resources—whales, fishes, forests, water, and the lands that could be harvested and grazed, and then traded overseas.

This chapter describes the impacts of large-scale dairy production on the watersheds of two large islands of Aotearoa. Before European settlement, the diverse Māori tribes or *iwi* celebrated the connections between ocean, river, and mountain and established a stewardship ethic called *kaitiakitanga*. Kaitiakitanga is an intermingling of beliefs, practices, and ethics that serve as a means to sustain resource use and place-based ecology and survival (Kawharu, 2000). Kaitiakitanga emphasizes the link between ecosystems and human well-being, and offers a more integrative approach to human resource use, place, and community. Kahui and Richards (2014, 6) note that kaitiakitanga is "the embodiment of a nexus of beliefs geared towards survival, and provides an integrated management system that bridges the gap between ecology and people." Kaitiakitanga represents a place-based sensibility, a way of reconnecting and restoring the human experience with the natural path of a river to the coast and sea.

RIVERS OF DIVERSITY

New Zealand is a country that receives abundant rainfall, and has more than 70 major river systems (30 on the North Island and 40 on the South Island). Of New Zealand's total length of rivers and streams, 51 percent lies in watersheds with predominantly natural land cover, such as native bush or alpine rock and tussock (Allibone et al. 2009). The remaining 49 percent of river length is in watersheds modified by agriculture (43 percent), plantation forestry (5 percent), or urban settlement (1 percent).

The watersheds and other aquatic habitats are essential to the maintenance of aquatic native species diversity. Many of New Zealand's animals and plants are not found elsewhere—these are known as endemic species. For example, over 80 percent of the 2,500 species of native conifers, flowering plants, and ferns are found nowhere else. Of the estimated 245 species of birds breeding in New Zealand before human arrival, 71 percent were endemic. This high proportion is mainly the result of the country's long isolation from other land masses. The best guess of the numbers of land-based native plants and animals is around 70,000 species. From human settlement to 1994, 43 (or 46 percent) of the 93 endemic land, freshwater, and coastal bird species became extinct, as did 4 of the 22 endemic seabird species (making 41 percent of all endemic species extinct; Rowan et al. 1997). Statistics New Zealand (2008) lists 2,788 native species and other taxonomic units (groups of organisms) as threatened.

The decline in native species indicates the poor health of the watersheds of the country (Allibone et al. 2009). New Zealand has drained 90 percent of its wetlands, felled over 70 percent of its native forests, and dammed, straightened, and re-engineered most of its rivers. Nearly half of the country's lakes and around 90 percent of the lowland rivers are now classified as *significantly* polluted. Nearly all national and regional river monitoring sites show declines in freshwater quality. A study by Bradshaw, Giam, and Sodhi (2010) that included an analysis combining measurements of countries' loss of native vegetation, native habitat, number of endangered species, and water quality found that, per capita, New Zealand was the 18th-worst of 189 nations when it came to preserving its ecosystems. The declines in the health of freshwater ecosystems are, for the most part, a result of poor land use and unsustainable agricultural practices.

As a consequence, New Zealand's wild and pastoral legacy is part nonfiction and part fiction. The country has spent decades establishing a brand, "100% Pure New Zealand," that remains a primary draw to tourists, who venture across the ocean to visit and explore its highcountry trails, fjords, rivers, and islands. The New Zealand brand builds on the glorification of the landscape's "pastoral design," with the façade of the small-scale sheep farmer roaming the steep mountains in a background of high-country, snow-covered alpine wilderness. The vision of the spectacular and seemingly untarnished natural backdrops of stunning waterscapes may support the brand, but it is a fading mirage. "100% Pure New Zealand" is predicated, in part, on a need to maintain the country's natural and pastoral heritage. But the major expansion and growth of large-scale dairy production across the two main islands of the country threatens this brand.

We can turn to the history of exploitation and colonization by Europeans to begin to peel back the green veneer of "100% Pure New

Zealand." New Zealand is very much part of a new global order driven by economic exploitation of resources that are traded overseas, to the profit of a few. Historian Leo Marx examined the challenge of protecting traditional values for the landscape in a socio-ecological context that has given rise to the mechanical, industrial age. In *The Machine in the Garden* (2000), Marx describes "two kingdoms of force"—the pastoral ideal, and the industrial modes of economic production and consumption. These two kingdoms of force represent a contested terrain of diverse epistemologies that shape social identity and worldview. The pastoral heritage is a mode of consciousness for Marx—not just a matter of expression but one of thought and behavior that shape relationships between human beings, society, and nature. Marx contends that the modern experience is one grounded in the tug of war between conflicting epistemologies of place. Human beings live in a "middle landscape" situated between nature's primal state (e.g., as represented by the myth of Eden) and mechanical civilization (e.g., as represented by the myth of progress and managerial control). Marx also describes the need for a "redemptive journey" (69) to a renewed ecological awareness and sensibility.

We can see these two kingdoms of force being played out in New Zealand in a number of ways. *Pākehā* is a Māori-language term for New Zealanders of European descent. It should not be understood as derogatory but as referring to the otherness and difference of Europeans in the context of Aotearoa. New Zealand is a country inhabited by Māori and *pākehā* (among others)—it is a country of 400 islands that is shaped by race, class, and diverse social and cultural affiliations and identities. Māori and *pākehā* can be understood as contested social identities that support diverse socio-ecological understandings of place and community.

The rivers of New Zealand are also caught in the socio-ecological context that includes these two kingdoms of force. A river is a lifeline that reinforces cultural identity and place-based knowledge. For Māori, the concept of *mauri* is the foundation of their view of resource use and watershed-based stewardship (Kahui and Richards 2014). The overarching goal of stewardship is based on the concept of kaitiakitanga. Kaitiakitanga emphasizes that the use and stewardship of the land and sea should maintain *mauri* to the point that the system is not changed or augmented by human activities. With respect to the value of *mauri*, harvesting resources from a river requires the maintenance of the ecological integrity of the system of relationships that supports the river's function and complexity. Kahui and Richards (2014) describe the sense of stewardship embraced by the Māori:

> This system is embodied in kaitiakitanga, which at a fundamental level incorporates a system of beliefs that pertains to the spiritual, environmental and human spheres. This provides the tools to manage the local environment; for example, spatial access and temporal access were regulated by *rahui* (reserved areas) and *owheo* (permanent conservation); maintenance of ecosystem structures were achieved by *ohu* (communal working bees) and rules pertaining to quantity and method of harvesting controlled extraction of resource units; metaphysical concepts such as *tapu* (involving restrictions of a spiritual nature) and *mauri* (vital essence; life force) enabled enforcement and monitoring of environmental indicators. Kaitiakitanga also included social protocols such as the ritual distribution of surplus by exchanging specialty foods from one area to another, usually both obligatory and reciprocal.

The river is an essential *life force* that connects a people, kinship, and community with the particularities of place. This concept of kaitiakitanga is similar to ideas held by several other Pacific island peoples. The Hawaiian concept of *mālama 'āina,* for instance, means caring for one's spiritual ancestor. The watershed and land are considered one's kin, an ancestor relative such as Papahanaumoku (Mother Earth) or an older brother, Haloanaka, who was born as a taro plant (Sponsel 2001). This indigenous spiritual ecology is a reflection of a deep ecological epistemology that unites place and society.

The river is also a machine to be used, managed, and engineered. The life force and ecological processes of the river system and greater watershed are transformed by modern society to support economic growth and instrumental values. As the river is managed as a resource, the cultural significance of the river's natural force (with its many offerings or services) becomes an issue of maximizing control and allocating resources well beyond the region or locale.

THE GLOBAL TECHNOLOGY COMPLEX

After centuries of colonization, many of New Zealand's major river systems now include the mechanical elements of hydro-dams, flood control, irrigation schemes, water diversion projects, and river control works. There are more than thirty-seven moderate and large (over 10 MW) hydroelectric power schemes in New Zealand. David Pietz (2002) refers to this process of transformation as a product of the "technology complex," which is based on the social alliances and networks of individuals that embrace the powerful mythologies of industrial capitalism. The fundamental characteristics of the technology complex are: the tools and structures of industrial technology; political organization; social mores and values in support of control; labor mobilization; other cultural symbols and mechanistic values (which are inherited or constructed); bureaucratic rationality; the commodification of the river; and the various modes of capitalistic development and growth. This technology complex, and more importantly the power it holds in society, contribute to the degradation and loss of free-flowing, wild river ecosystems. As the power of the technology complex emerges and takes hold, the ecological integrity and the social identity of the peoples who inhabit watersheds change.

The growth of the technology complex was an important part of the early development of industrialization. The eighteenth and nineteenth centuries in Europe, the United States, and China were periods of dramatic change for major river systems. During the Ming Dynasty, in the middle of the 1800s, the great Yellow River and its floodplain were constructed into a major agricultural area, with the development of major irrigation systems and water diversion projects. The contemporary manifestation of the denaturing effects of hydropower development and the illusion of flood control is China's Three Gorges Dam.

In the western United States, Donald Worster (1992) documents the natural history of the emergence of an empire that with technology redirected the river systems for agriculture, transport, flood control, water, and energy. New energy-sheds and food-sheds fed industrialization and a growing population across the West, and generated great wealth in the hands of a few. Richard White (1995) describes one consequence of the transformation of rivers as the birth of the "organic machine"—part human-made mechanical artifice and part natural system. White describes the transformation of the Columbia River, in the northwest part of the United States, as "one unified machine, one organic whole." White does not leave us with much hope. The river remains uncontrollable despite our wishes to contain and master it for our economic ends. Human beings do not triumph over nature, nor does nature triumph over us. Rather, White describes the worldview at play in transforming the river, and he reminds us that when we turn natural areas into property or a mere commodity we have let go of any semblance of the social or ecological values that are essential to our well-being. In viewing the Columbia as a machine we denatured nature, and now we have less nature to draw from. The social construction of nature is shaped by the values of a mechanistic sensibility. It is the mechanical force that reimagines and reconstructs the river-as-machine to be turned off or on like a house faucet.

Dams often impound the awe of the river. The dam makes the river flow dependably and the riparian landscape safe to inhabit; plus, the electricity from the dam makes life comfortable. Nevertheless, the awe of the spring runoff is gone. The "river" exists in some type of middle ground between its mechanistic and natural characteristics. The river is part nature. But nature is beyond our capacity to control or manage. As we lose the natural features of the river, human society becomes more ecologically vulnerable as dramatic declines in availability of and access to clear water continue across the globe. This is a not-so-subtle consequence of our treatment of rivers and watersheds.

Like so many other places in the world, New Zealand's waterways exist between two kingdoms of force. Ironically, water scarcity is becoming manifest locally across New Zealand. There are two dimensions to the insecurity that New Zealand has constructed for itself. First, research on "virtual water" relationships between consumption, trade, and water resource use shows that freshwater resources should not merely be considered as an issue for individual countries or river basins. Exporting dairy and other animal products requires a significant amount of water. On average it takes 1,020 liters of water to produce 1 liter of milk (Hoekstra 2012). Overall, the international virtual water flows related to the global trade of dairy and other animal products add up to 272 billion m³ per year, which is the equivalent of about half the annual flow from the Mississippi River in the United States (Mekonnen and Hoekstra 2011). Hoekstra (2012, 7) writes:

> Until today, water is still mostly considered a local or regional resource, to be managed preferably at the catchment or river basin level. However, this approach obscures the fact that many water problems are related to remote consumption elsewhere. . . . Water scarcity is not translated into costs to either producers or consumers; as a result, there are many places where water resources are depleted or polluted, with producers and consumers along the supply chain benefitting at the cost of local communities and ecosystems. It is unlikely that consumption and trade are sustainable if they are accompanied by water depletion or pollution somewhere along the supply chain.

Second, climate change is influencing the rainfall in New Zealand. In 2013, the country faced its worst drought in thirty years. Parts of New Zealand's North Island, where most of the dairy is produced, are drier than they've been in seventy years. So, water scarcity is an economic reality, and the large quantity of water required to feed grass and dairy cattle impairs the country's ability to adapt to climate disturbance. In 2013, $820 million in lost export earnings in dairy production resulted from the drought. The drought also required ranchers to reduce the herds for dairy.

COMING INTO THE WASTESHED

New Zealand is failing to live up to its brand, which advertises a clean, green country (Anderson 2012). Watersheds are now *wastesheds*: systems of shedding waste and other by-products of industrialism. The wastesheds of the country are the subject of a heated political debate that includes increasing international attention on the failure of the "100% Pure New Zealand" promise (Harding 2007). As one freshwater scientist (Joy 2011) observes: "We have gone too far. Surely it is time to admit, even if just to ourselves, that far from being 100% Pure, natural, clean, or even green, the real truth is we are an environmental /biodiversity catastrophe. . . . In five decades New Zealand has gone from a world-famous clean, green paradise to an ecologically compromised island nation near the bottom of the heap of so-called developed countries." More than 60 percent of New Zealand's native freshwater fish (among them the only freshwater crayfish and mussel species) are listed as threatened with extinction. A 2009 study ranked 34 of 51 (or 67 percent) of native freshwater fish taxa as threatened or at risk (Allibone et al. 2009).

Responding to the pressures associated with the wastershed requires the application of scientific information in future planning, and a renewed sensibility that can protect ecosystems and place. In the early 1990s, New Zealand was considered to be in the forefront of watershed-based planning and policymaking in the world; the country's watershed-based planning program was considered by scholars an ecologically progressive and green ideal for other countries to emulate. Unitary systems of governance were created that reflected the complexity of the land–sea interface. Regional and district councils were empowered to manage freshwater and marine systems. New Zealand is one of the few countries in the world that encourage the development of a planning and regulatory permitting framework that includes a major watershed and associated coastal marine ecosystems that extend out to twelve nautical miles (i.e., the territorial sea). Few countries have developed a political jurisdiction that links a watershed with marine ecosystems. But despite the progressive character of the country's watershed-based approach, New Zealand is in a new era of economic development and trade that cannot support or maintain ecosystems.

Since New Zealand is part of the Commonwealth, the Crown exercises control of freshwater, largely through the Resources Management Act of 1991 (RMA), which delegates watershed planning primarily to regional councils. The RMA was a product of policy and institutional innovation that began in the late 1980s. The reforms were driven by a growing free-market ideology, the widespread desire to shrink central government, and an overly complex and prescriptive regulatory system. An extensive stakeholder consultation effort led to an unprecedented alignment among business, government, and the public interest community in support of the reforms. More than 800 governmental and quasi-governmental agencies were dismantled or reorganized. In their place, 3 primary central government agencies and 86 local government authorities (12 regional councils based on catchment boundaries, and 74 territorial authorities called district or city councils) were established, which would be collectively responsible for all aspects of environmental, natural resource, and land-use planning and management. In addition, over 55 statutes and 19 sets of regulations were eliminated and replaced by a single legislative enactment—the RMA—encompassing environment, natural resources, and land use beneath one umbrella for the purpose of promoting sustainability (Makgill and Rennie 2012).

Soon after the passage of the RMA in 1991, the country created fifteen regional councils. Regional councils are responsible for controlling the taking, use, damming, diversion, and pollution of freshwater (RMA, section 30 (1), (e) and (f)). The regional councils have authority to set and decide on flow regimes, and to manage and allocate water and activities in riverbeds. Makgill and Rennie (2012, 149) describe the importance of the RMA:

> The definition of environment under the RMA of 1991 includes ecosystems. That means that a proposed activity must also be considered in terms of its adverse effects on an ecosystem under section 5(2)(c). The ecosystem approach is also present in the division of regional councils' spheres of authority into catchments. Catchments are the ecological conduit for the passage of water to the coast. Understanding land-water relationships is the starting point for appreciating the need for holistic management of regions using natural boundaries. The primary determinant of the health of any near-shore marine ecosystem is run-off from contributory catchments. Chemical contamination from run-off results in the overfeeding and,

frequently, the poisoning of estuaries. Regional councils . . . have responsibility under the RMA for the integrated management of natural and physical resources within their regions. Their regions include both the land catchment and the offshore environment out to the 12-nautical-mile territorial sea boundary.

The use of scientific information is an essential component of decision-making and regulatory planning under the RMA. However, local governments and regional councils have "increasingly departed from the statutory process" and have turned toward a more cooperative approach to industry and other user groups in policymaking (Bremer and Glavovic 2013, 107). Large-scale dairy interests have a significant level of power in watershed-based planning and decision-making. The country has adopted a *co-management* approach to resource planning that includes new collaborative relationships between industry and policymakers.

One reason for the shift to dairy is an increasing demand for milk products in Asia. New Zealand provides roughly $4 billion per year in dairy products to China (as of 2013). (To put this in perspective, the total value of agriculture productivity in the Salinas Valley in central California, known as the "salad bowl of the world," is about $4.2 billion.) Roughly 37 percent of the country is now used for dairy (DairyNZ 2012). The demand for dairy in Asian markets is increasing, and so too is the value of milk powder as an export. Due to public perception in China of the poor quality of their own milk supplies, they have looked overseas for dairy and protein-rich animal products. In addition, with the loss of their primary protein sources from the sea, Asian countries have shifted from fish to beef and dairy imports. As many of China's citizens have become wealthier, the average daily intake of animal-source foods in the country has more than tripled; and this has led to increased risk of heart disease, diabetes, and colon cancer in the country (Popkin and Du 2003).

Dairy production is also not without great socio-ecological costs to New Zealand's watersheds. Thirty-seven percent of the dairy cows are on the South Island (DairyNZ 2012). The majority of dairy herds (76 percent) are located on the North Island, with the greatest concentration (30 percent) in the Waikato region. The Taranaki region, with 15 percent of the dairy herds, is the next-largest region on a herd basis. Twenty-five percent of all dairy cows are located in the Waikato region, followed by North Canterbury (12 percent), Southland (11 percent), and Taranaki (10 percent).

New Zealand's Dairy Production

- Between 1990 and 2010, the national dairy cattle herd increased from 3.4 million to 5.9 million; and the use of nitrogen fertilizer increased by more than 800 percent (the highest percentage increase across 29 OECD countries). Phosphate fertilizer use increased by more than 100 percent (the second-highest increase in the OECD).
- The reduction in beef cattle and sheep numbers, and any positive consequences for water quality that this might have had, is outweighed by the large increase in dairy cattle.
- The most intensive farming is happening close to water sources, and in areas of groundwater recharge. The vast majority of dairy production (92 percent) is supported by Fonterra, a multinational dairy producer.
- Major dairy export markets are China, with an 18-percent share, the Philippines (4 percent), Algeria (4 percent), Australia (4 percent), and Saudi Arabia (4 percent), as of 2011. New Zealand dairy farms processed 17.3 billion liters of milk in 2011. New Zealand produces approximately 2 percent of total world production of milk, at around 16 billion liters per year. The main dairy exports are concentrated milk, at 58 percent, butter (21 percent), cheese (11 percent), whey and milk products (6 percent), and not-concentrated milk (2 percent), as of 2011. Approximately 95 percent of all New Zealand dairy production is exported.
- Dairy production has increased by 77 percent during the past twenty years in New Zealand.

While the New Zealand dairy industry is providing dry milk powder to China, important consequences for the country's watersheds are worth expanding on. There is increasing public opposition to dairy production. The pastoral heritage and the natural integrity of the country are at stake. The small family farm has given way to large-scale dairy production. A great majority of the dairy production in New Zealand is by Fonterra, which is a multinational company, cooperatively owned by 13,000 New Zealanders, who together produce 22 billion liters of milk each year. New Zealand leads the world when it comes to dairy (see box), accounting for over a third of the world's international dairy trade. The country's dairy products feed more than 100 million people worldwide, and the dairy industry contributes 25 percent of New Zealand's merchandise export earnings (DairyNZ 2012). The shift to large-scale dairy production has led to an overall increase in the number of dairy cows per hectare in the country, and a general increase in the amount of land owned and operated per farmer.

The Ministry for the Environment (2012) describes a number of watershed-related problems that are, in part, based on pressures and impacts from the dairy industry, including:

- Few catchment-level water quality standards are currently set.
- Managing water quality, especially diffuse sources, is complex and creates difficulties for all stakeholders.
- Lowland rivers in agriculturally developed areas have been subjected to high nutrients, turbidity, and fecal contamination, leaving them in a poor condition.
- Streams in areas of dairy farming, especially where poor practices of shed effluent disposal have been used, are in particularly poor condition, and the intensification of farming associated with dairying in general has also been related to increasing levels of nutrients, sediments, and fecal bacteria.

Also observed are:

- alteration and destruction of habitats and ecosystems
- widespread and increased eutrophication
- decline of fish stocks and other renewable resources
- changes in sediment flow due to hydrological modification.

In the Ministry for the Environment's (2012) report card for water quality, most rivers and creeks had "poor" or "very poor" quality (52 percent of monitored river sites). A further 28 percent were graded "fair"—with a risk of illness for those swimming there. Only 20 percent of monitored river recreation sites were graded "good" or "very good." Recent studies show that water quality and quantity are on the decline in New Zealand, primarily due to dairy production (Morrison et al. 2009). Data from the National River Water Quality Network also shows that water quality in New Zealand is declining. In particular, there is a significant increasing trend in nitrogen (as nitrate and nitrite), dissolved reactive phosphorus, and total phosphorus levels. In addition, the water take has almost doubled in the past ten years, and at the same time water quality has fallen. Freshwater surveys show that many places are not safe for swimming.

There are a range of externalities associated with dairy production, including soil compaction, nitrate contamination of drinking water, nutrient pollution, and greenhouse gas emissions. Externalities are the economic and ecological costs that are rarely part of the production value of a product like milk or meat. For instance, the amount of water it takes to produce each liter of milk is not part of the price we pay for milk. The costs or "externalities" associated with water and air pollution is also not part of the price consumers pay for dairy products. A recent study assesses the economic costs of these ecological externalities (based on a 2012 evaluation) and finds that these costs surpassed dairy's contribution to the country's gross domestic product in 2010 (Foote, Joy, and Death 2015). In other words, the cost of large-scale dairy production to the ecosystems of the country exceeds the total economic value and net worth received from dairy exports. This is an important finding because the ecological costs of agriculture land-use activities are rarely compared to the economic benefits.

The green veneer of the New Zealand dairy industry's brand is slowly fading. This green brand is key part of the marketing of products from the country. The country's dairy productsare marketed as being produced in a sustainable way and in a natural landscape. As the green veneer is exfoliated by evidence from the scientific community, a political debate has erupted over the activities and land-use impacts of the industry. The residents of New Zealand have begun to question their own identities. In an interview by the BBC in January 2012, the country's prime minister, John Key, compared New Zealand's "100% Pure" tourism marketing campaign to a fast-food ad. "It's like saying 'McDonald's, I'm lovin' it'—I'm not sure every moment that someone's eating McDonald's, they're loving it. . . . It's the same thing with 100% Pure," he said. "It's got to be taken with a bit of a pinch of salt."

THE JOURNEY AHEAD

A "redemptive journey" is one response to the recognition that the watersheds of the country are being transformed into wastesheds. A redemptive journey to home place, or what I earlier referred to as a homecoming, is a process that will require negotiating with others to support the roar of a river, an ecological understanding of the true costs of global economic trade and exploitation, and respectful care for the range of species that are threatened by unsustainable land-use activities, such as dairy production. The recognition of degraded watersheds can be a motivating force behind a stronger commitment to one's place in the world, and can contribute to a new shared vision that protects and restores the more-than-human community. A renewed place-based sensibility is needed across the bioregions of New Zealand.

The need to restore the tuna or eel to the watersheds of New Zealand is an example of an ecological sensibility that combines and reconnects the tuna, place, and *iwi* (and Kiwi). In this sense, the tuna is the memory of a river lost to the world; the tuna is a ghostly memory that exists underneath the river of the mechanistic world. As chapter 3 showed, the memory of wild rivers and the loss of salmon are the primary motivating factors that continue to foster California's watershed-based movement. The biophysical tipping point that indicates the loss of a river ecosystem can, in time, encourage a social turning point and the creation of place-based movements and new social alliances and networks. A greater ecological awareness of one's place can start with identification of one's wasteshed.

A strong literary tradition in New Zealand supports the reconnection of people to place, rivers, and watersheds. A sense of place-based community can encourage respect for the circle of animals, plants, and insects that shape social identity and a more civic-minded cultural practice. Indigenous and colonial histories of New Zealand's watersheds are described in David Young's *Woven by Water* and Geoff Park's *Nga Uruora: Ecology and History in a New Zealand Landscape*. Young provides stories that inspire a reconnection of diverse peoples and places of the Whanganui River, the most distinctively Māori catchment in Aotearoa. This remarkable book begins in the early 1800s and traces oral and archival stories of the Whanganui River.

Park is referred to as New Zealand's Gary Snyder, and has inspired a new generation of ecological activists. He writes: "New Zealand entered the European imagination as a land where they could freely expand their preferred way in the world. . . . Led by beliefs as ancient as the Roman Cicero's—we are the absolute masters of what the earth produces. The rivers are ours. We stop, direct, and turn the rivers . . . to make it, as it were, another nature" (1995, 14–15). Park describes what has been lost since Europeans arrived to cultivate and use the landscapes and watersheds of New Zealand. He also calls for the cultivation of a new social identity for Kiwis—toward an ecological sensibility that embraces "ways of knowing *with* country" (my emphasis). This sense of knowing "with country" is a reference to the cultivation of a place-based knowledge system that is supportive of a deep interdependence between the human and more-than-human community characteristics of a region and home. Park is calling for a return to the values of kaitiakitanga and a more intimate way of interacting with one another and the ecology of a place.

Other works by European writers about exploration in the country situate rivers as almost sentient antagonists. Rivers are described as very angry and threatening. The theme of the power of water runs right through the poetry of James Baxter's "Poem in the Matukituki Valley." There is also writing about sustenance from, linkages with, and ordinary life in relation to water. Ian Wedde's poem "Pathway to the Sea" describes someone who builds a drain to stop rain carrying sewage into the sea. Much of this literature calls on a new identity to take hold—one that embraces home place, the river, and community.

Hone Tuwhare, a poet of Māori decent, reflects on this struggle to maintain cultural identity in the face of large-scale degradation of the country's ecosystems. Tuwhare's poetry often engages rivers, speaks with them directly, and emphasizes that they also express themselves, as depicted in his poems "The River is an Island" and "Deep River Talk." In his poem "Not by Wind Ravaged," Tuwhare (1964, 20) writes:

> Deep scarred not by wind ravaged nor rain nor the brawling stream:
> stripped of all save the brief finery of gorse and broom;
> and standing sentinel to your bleak loneliness the tussock grass
> — O voiceless land, let me echo your desolation. The mana of my
>   house has fled, the marae is but a paddock of thistle.

Marae is the Māori term for a sacred place, while *mana* is the magical power often associated with these special places. The knowledge or epistemology that grounds social identity is a common aspect of those *iwi* or tribes of the Māori that respect and care for the land. On the other hand, it is important not to romanticize the past treatment of nature by the Māori; the *iwi* overhunted and exploited the landscape, and their conflicts and wars impacted the natural features of the landscapes they inhabit.

THE SACRED ECOLOGY OF KAITIAKITANGA

Notwithstanding the impacts of native peoples on the landscape of New Zealand, acknowledging the multiple values carried by traditional ecological knowledge, or what Fikret Berkes (2012) describes as the "sacred ecology" of indigenous peoples, is one step toward a redemptive journey that supports a cultural sensibility of home place. Sacred ecology is a knowledge-practice-belief system and includes four interrelated levels: local knowledge that is place-specific; resource stewardship that includes the integration of knowledge with practice; social institutions that include formal and informal rules and codes of behavior; and worldview and belief structures.

The Māori concept of kaitiakitanga is an example of sacred ecology. Kaitiakitanga is an essential part of the custom and practice of the *tangata whenua* (Māori for "people of the land") (Roberts et al. 1995; Kawharu 1998; Ruru 2009; Selby, Moore, and Mulholland 2010). The Māori embrace a concept of *whakapapa* (genealogy) which expands the sense of kinship so that family extends to the mountain, river, coastal, and marine area. Māori have passed on the values of *whakapapa* and kaitiakitanga across generations in story, mythology, and practice. These traditional values denote the authority for the exercise of the stewardship obligation as deriving from the *atua* (ancestors). In this way, the use of natural resources is not without limits; conservation and preservation are important features of resource use for traditional Māori society.

Kaitiakitanga had inspired a range of conservation and restoration efforts throughout New Zealand. The multifaceted epistemology and traditional value of kaitiakitanga have been institutionalized in a number of treaties, such as the Treaty of Waitangi (1840), and related settlement statutes, and laws that are the foundation for relationships between diverse Māori and the British Crown, including the RMA (Kahui and Richards 2014). The spiritual component of traditional ecological knowledge acknowledges the influential powers in all things.

On the other hand, the value of kaitiakitanga remains a major source of conflict, given the English and Māori interpretations of this epistemology and how to put it into practice. In many ways, the struggle to support the concept of kaitiakitanga in the face of industrialization and overuse of the land and water remains a social and political facet of the challenge of maintaining Māori identity. Awareness of the coastal area where the tribe's canoe landed, the path travelled upriver, the direction of the wind, an understanding of where to gather food and sustenance, and the spiritual connection to the mountain and native bush is perpetuated across generations by the Māori (and Kiwis) who share the value of kaitiakitanga.

Although the concept of kaitiakitanga is embodied in the treaties and legal framework of the country, it is not part of the socio-ecological context of the *pākehā*. Unlike the European settlers, the sense of belonging to a place represented by cultural values such as kaitiakitanga broadens the sense of *iwi* kinship to *rivertime*. This notion of rivertime invokes the sense of a flowing river over time, the changes in the ecology of the river over time, and how the river changes the communities that depend on it for their economies and lifestyles. This notion of rivertime is supportive of a cultural heritage that the *iwi* are born into: it is a right and obligation of stewardship that is inherited across generations, and the knowledge of the river and greater watershed is an essential feature for the sustainable use of the river across time. The fact is that several *iwi* of the Māori are major landowners, owning major quotas in commercial fisheries (roughly 40 percent) and aquaculture (roughly 20 percent) and owning the lands that are used for major industrial-scale agriculture. The spiritual component of kaitiakitanga may simply become a fading reference point to a past worldview. As nature is transformed by Māori and *pākehā* commercial enterprises into a mere commodity, kaitiakitanga is threatened.

THE RETURN OF THE EEL

As a totem species, the New Zealand longfin eel is a galvanizing force that combines diverse elements of the natural and human worlds. In Tuwhare's poem *A Tail for Maui's Wife* (1974, 43), he describes the importance of the eel: "I move with her I move against her I move inside her / She is water." The eel is a primary emblem in Māori mythology. In an essay on the importance of eel to the Māori, James Prosek (2010) writes: "A girl is cleaning her clothes in a spring-fed pool and is violated by an eel that was once her beloved pet. A villager captures the eel and cuts its head off, and the girl buries the eel in the sand. The first coconut tree then grows from it."

The eel brings together the *iwi*, their respective places, home, and region—it connects the people to a common watershed that extends beyond the coast to the deep sea. In this sense, the eel embodies the irrevocable interdependence of the people and nature, and the terrestrial, river, and marine ecosystems of New Zealand and the greater South Pacific. The eel remains an essential facet of celebration, and a primary source of food for ritual and other cultural activities for the *iwi*.

The eel is a species endemic to New Zealand, with a very wide distribution through New Zealand's freshwater waterways, including the Chatham Islands. The eel's life cycle depends on both freshwater and saltwater habitats. It is believed to spawn at sea (perhaps near the Cook Islands). The species is listed as threatened by New Zealand Department of Conservation because of overfishing, hydrological modification, habitat loss, and pollution. Trap and transfer operations at some hydro-dam sites in recent years have revealed that the number of longfin eels moving up rivers is very low—at least a 75-percent reduction—in stark contrast to the huge runs that were witnessed prior to the 1960s (Graynoth 2006). Commercial catch records reveal a trend of decreasing size of all eels caught, most (in 2007, 96 percent in the heavily fished Waikato River, 50 percent nationwide) now being within the lowest size category (220–500 g). Since the early 1990s the commercial harvest of eels has halved due to this rapidly declining population. Another key issue is habitat loss due to land-use changes and development, including wetland drainage, the construction of dams, irrigation schemes, river diversions, and culverts. Hydro-dams are a particular problem as they interfere with the migration pathways of eels. One estimate suggests that hydroelectric dams have blocked the longfin eel's access to the sea in 35 percent of its habitat.

The call to protect the eel is a response to the plight of the species and greater watershed. The struggle to live life in accordance to kaitiakitanga remains. This struggle is also reflected in the fundamental challenge facing the diverse peoples of New Zealand, who are caught between two kingdoms of force. While the external demands for resources exist, as exemplified by the demand for dry milk, beef protein, and other dairy products in Asia, the redemptive journey requires a powerful commitment to act ecologically on behalf of place.

The traditional method of greeting in Māori is the *hongi,* which is completed by pressing one's nose and forehead (at the same time) to another person's at an encounter. It is used at traditional meetings among Māori people and in major ceremonies. The power of the greeting resonates in the sharing of a common breath. This common breath extends to the river, the lifeline that connects people and places. In the *hongi,* the *ha* (breath of life) is exchanged and intermingled. In some ways it is similar to the Polynesian greeting of *aloha,* which can be translated as "of the breath." With each greeting, you are reminded of your obligation to share in all the duties and responsibilities of the home people.

CHAPTER 5

Organic Machines and the End of Offshore Oil

An organic system directs itself to qualitative richness,
amplitude, spaciousness, free from quantitative pressure and
crowding, since self-regulation, self-correction, and self-pro-
pulsion are as much an integral property of organisms as
nutrition, reproduction, growth, and repair.

—Lewis Mumford (1961)

You get to know the ecosystem you depend on by interacting and engaging with it. From afar, the complexity of this ocean planet is difficult to
completely perceive and comprehend. Few understand and interact
with the diversity of marine life that lies underneath the blue horizon—
an abundant and diverse ocean of life that is well beyond our capacity
to see, smell, or hear. In a mechanical age like ours, we can lose sight of
the importance of the communal norms of reciprocity and responsibility that are needed to care for the ocean. It is not a limitless frontier for
our waste, trash, plastics or pollution. Nor is the ocean a mere resource
for us to use and exploit.

As a child, I grew up near coastal oil development. Oil development
is a mechanical leviathan that is perpetuated by our thirst for fossil fuel.
Oil and other fossil fuels have long fostered the greed and growth
endemic to our global economy. Where I played during the summer,
I could smell the oil along the upland and soft edge of the Bolsa Chica
wetlands in Huntington Beach, California. My feet would be covered in
the goo. I learned that oil removes oil, but it is hard to remove the oil
from the sea or coast.

For thousands of years, the diverse prehistoric Chumash peoples of
Southern California used the natural seepage of oil that washes up on

the shore to line their canoes. These canoes were used to travel from the coastal mainland out to the Channel Islands. Natural oil seeps occur in many areas in Southern California.

Later in the eighteenth century, off the shore of California, the first "oil wells" were whales killed by harpoons; whale oil was the primary source of heating for early industrialization. Before the Europeans came, the bays and inlets of the state were teeming with whales. You could smell the whales gathering in Monterey Bay, for instance, far inland. Whales were present in the wetlands of Los Angeles and the Santa Monica Bay. While our relationship to whales near California has changed, our relationship to oil as the primary source of energy has not.

The growth of mechanical society became dependent on a second source of energy. Interest in oil seeps became widespread after the 1859 discovery of oil in Pennsylvania, when the value of kerosene as an illuminant became generally known. Thousands of wells were developed in the late nineteenth century along the coast and in the canyons or within the watersheds of Southern California and Texas. Upton Sinclair's novel *Oil*, published in 1927, depicts the oil boom that contributed to corruption and greed in the hinterlands of Southern California.

This same greed contributes to the exploration for and development of offshore oil and its associated wastes. In time, with changes in technology and increased demand for fossil fuels, short piers from the coast were built to support the country's first offshore rigs. Hundreds of piers were constructed in Summerland, just south of the city of Santa Barbara, and other places in the south coast of California.

As oil development moved farther offshore, federal, state, and local authorities began to fight over political jurisdiction for governance and a share in the wealth produced by offshore oil fields. There were disputes over the economic royalties that many government authorities maintained should be paid to the state. There were disputes over local conservation areas that were established by municipalities that prevented oil development. The ownership of the oceanic commons was an issue of conflict and debate. This conflict over oil profit and authority in the United States led to the passage of the Outer Continental Shelf Lands Act (OCSLA) in 1953. OCSLA defines the outer continental shelf (OCS) as all submerged lands lying seaward of state coastal waters (3 miles offshore) which are under U.S. jurisdiction. Under OCSLA, the secretary of the interior is responsible for the administration of mineral exploration and development of the OCS. The act empowers the secretary to grant leases to the highest bidder on the basis of sealed competitive bids, and to formulate regulations as necessary to carry out the provisions of the act. The act, as amended, provides guidelines for implementing an OCS oil and gas leasing program to support the exploration, development, and removal of offshore oil structures. Today, California's jurisdiction extends from the shoreline out to 3 nautical miles offshore, while the federal government claims jurisdiction from 3 to 200 miles offshore (this is known as the exclusive economic zone).

Conflict over proposed offshore oil and gas exploration and development did not end with the passage of OCSLA. Oil development remains a fierce issue of contention between states and the federal government, and between conservationists and the oil industry, among other stakeholders. One reason for these conflicts is the impacts of offshore oil exploration and development. In offshore oil exploration, oil companies use sonar or acoustic soundings to detect oil beneath the sea floor. Marine mammals are sensitive to sonar and marine noise. Also, over time the offshore rigs and their associated pipelines and structures lose their economic value. During development of an offshore oil lease, there comes a time when the economic cost of removing remnant oil is not worth the industry's effort. Removal options are considered under the general concept of decommissioning of offshore oil structures.

This chapter reviews the roles of science and values in the decommissioning of offshore oil structures in the Gulf of Mexico and the marine waters off Southern California. The Gulf of Mexico contains approximately two-thirds of the world's offshore oil and gas structures, and 95 percent of the production platforms in the exclusive economic zone of the United States (Reggio and Kasprzak 1991, 11). The Gulf rigs are the planet's leading source of methane gas, a major contributor to the greenhouse effect, ocean warming, ocean acidification, the melting of the polar ice, sea level rise, and general climate disturbance. Approximately 100 rigs are removed from the Gulf waters each year. California's offshore oil and gas drilling rigs are located in both state and federal waters, with twenty-three of the twenty-seven active platforms in federal waters. There are a number of alternatives to complete removal of these offshore structures, including partial removal, leaving the rig in place, and potentially "reefing" the structure in one form or another in a different marine area. Most of these offshore structures will be decommissioned within two decades of now. The question is, what will be left behind after the end of offshore oil development?

Removal of offshore oil structures is complicated because there are ecological and economic factors to consider. This chapter provides a profile of the ecology of offshore oil structures in the distinct contexts of the Gulf and Southern California. Values and science shape the politics of decommissioning these structures (McGinnis 1998; McGinnis, Fernandez, and Pomeroy 2001; Carr et al. 2004). Many structures include tons of biomass that has accumulated in the submerged portion of the rig. In the Gulf, there are approximately 5,100 offshore rigs, which have made areas off the shore of Alabama one of the top commercial fishing areas of the country. Submerged structures like offshore oil rigs can provide habitat for fished species, but this depends on the ecological and oceanographic conditions that are associated with the submerged structures. A few of the twenty-seven oil rigs and structures off Southern California are three times the size of the Empire State Building and include over 75,000 tons of scrap metal and organic material. The National Research Council (1996) has estimated that the cumulative costs for removal of all platforms in the OCS could total $9.9 billion by 2020. There are economic benefits to the industry in pursuing partial rather than complete removal; the cost savings would be significant. According to some industry estimates it can take hundreds of millions of dollars to completely remove a large offshore oil structure. States are also interested in receiving part of the cost savings that would occur in a partial-removal option. Given these economic factors, including the potential benefits rigs have in enhancing fish populations, the coastal states of the Gulf of Mexico adopted laws in the late 1990s that supported a rigs-to-reefs alternative to complete removal. In 2010, California created public policy that allows consideration of partial removal for a rig on a case-by-case basis.

CATASTROPHIC OIL SPILLS

In June 2015 a pipeline that transported oil from an offshore rig ruptured at Refugia State Beach in Santa Barbara County, California. The oil soon created a slick in the near-shore marine system, and oil covered the coastal region. This stretch of coast is close to my heart. It is my home that was covered in oil. I spent twelve years working with landowners, farmers, ranchers, and conservationists in a collaborative process to protect this stretch of coast, which the community refers to as the Gaviota. It is truly one of the hot spots for threatened biodiversity in the world, where central and Southern California marine, coastal, and terrestrial species and habitats meet in a very important transition zone, or ecotone. The Gaviota Coast has a rich history of cultural inhabitation by the coastal band of the Chumash peoples, who inhabited the region for roughly 8,000 years. The oil slick migrated down the coast, killing dolphins, suffocating fishes, depriving invertebrates of oxygen, and drenching the feathers of shore and seabirds. Brown pelicans covered in oil suffered greatly. One consequence of the oil slick is that ExxonMobil shut down three platforms off Santa Barbara and closed its oil processing facility at Los Flores Canyon along the Gaviota Coast.

This is the history of offshore oil development—it includes the devastating social, economic, and ecological impacts of oil catastrophes. Each spill is beyond our control to manage its impacts. The oil slick reminded the nation of the devastating oil spill in the Santa Barbara Channel in 1969 that in many ways contributed to the passage of the country's major environmental laws. It also showed that oil spills jeopardize marine protected areas. Three important marine reserve areas were impacted by the oil spill in the near-shore system along Gaviota. For a short time, oil spills are newsworthy, and for several weeks stories on the impacts of oil on the Gaviota Coast were reported on by the international and national media. But such stories rarely reflect the direct human and ecological impacts of these so-called "spills." The coastal province of Gaviota is sacred to me. I have spent countless hours on the region's beaches, and thousands of hours working with community members to try to protect it. This oil catastrophe is close to me; I smell it, and feel its consequences. My heart sank with the sight of so much suffering, of life smothered by the oil, my sacred place tarnished and destroyed.

The history of oil-related catastrophes (figure 5.1) is a primary reason why conflict over offshore oil development and decommissioning activity continues today. There have been globally publicized catastrophic oil spills in the marine waters of the United States—Santa Barbara (1969), *Exxon Valdez* (1989), and the Gulf of Mexico (2010). Few of these spills led to the types of policy innovation that the 1969 oil spill had at the federal level. Yet, local citizens along the California coast have, in general, remained adamantly opposed to offshore oil development. There have also been "silent spills" that have not received public and governmental attention, such as the spill of millions of gallons of oil in the aquifers under the Guadalupe Dunes on the central California coast (Beamish 2002). This spill contaminated the aquifer under the dunes that provided water to the residents of Avila Beach. It was over eighteen years before a whistleblower from the oil company told the story of the spill. Government agencies, the industry, and other elected

Largest Spill
First Gulf War
Kuwait
1991
5,700,000

Largest Spill from a Well
Ixtoc 1
Mexico
1979
3,300,000

Largest US Spill
Deepwater Horizon
Gulf of Mexico
May 21–30, 2010 (ongoing)
504,000–798,000*

Largest Previous US Spill
Exxon *Valdez*
Alaska, USA
1989
206,000

*12,000–19,000 barrels per day
Source: Oil Spill Intelligence Report; all figures are estimates

FIGURE 5.1 Ocean oil spills (all numbers in barrels).

officials knew of the spill, yet did not take action. These spills were not accidents per se but the consequence of gross negligence on the part of the oil industry that was allowed to occur in the absence of responsible federal government oversight. The history of offshore oil development also reveals a cozy relationship between the oil industry and those government officials who are responsible for managing the industry. In addition, the culprit for these blowouts and catastrophes is more widespread than we would like to believe—we all share responsibility because we have had too much faith in the technology to control and manage offshore oil, and because we remain blinded by our culture's dependence on fossil fuels to support our economies. There is no technology that can control and contain the devastating impacts of oil spills.

There have been many other spills, less publicized and as ecologically wasteful and culturally ravaging as the major spills covered by the media. They all share a common story about the ways human beings relate to and abuse the ocean. More often than not the social and ecological impacts of oil-related catastrophes are underestimated, and the recovery rate of the ecosystems and the communities affected by the spill is an issue of scientific debate. In addition, the studies funded by the oil industry that examine the biophysical impacts of oil catastrophes are often not accessible, and remain the "property" of the industry. There are also few long-term social and economic studies of the impacts of spills on local communities, who may suffer for generations from the loss of essential resources. The cultural ramifications of spills vary considerably, but particular places feel the long-term impacts of degraded ecosystems and lost resources even if the news media fail to report it. Riki Ott (2008) reflects on her experience in the small fishing town of Cordova, Alaska, which was forced to respond to the social, economic, and emotional trauma of the *Exxon Valdez* spill in the Prince William Sound. She writes, "These disasters so thoroughly disrupted people's lives that entire communities became dysfunctional" (xv). In Cordova, one result was that the people of the places so devastated by these disasters developed new forms of trust and awareness, and a stronger sense of community emerged. They began to rebuild the community in the face of the catastrophe. No less is needed today to confront the pressures, threats, and impacts from the slow violence of climate change.

Catastrophe can be a catalyst that binds people together for a common cause and a shared vision of a new future. In the wake of Unocal's Platform A spill in the Santa Barbara Channel in 1969, a strong network of newly created conservation and legal organizations, led by the Environmental Defense Center, and anti-oil organizations, such as Get Out Oil (or GOO), was galvanized and contributed to a resurgence and growth of the contemporary environmental movement across the United States. This coalition included groups like Save the Bay, which was established in 1962 to prevent the use of the San Francisco Bay as the primary landfill for the city. The 1969 spill was well publicized, and was part of the inspiration that led Senator Gaylord Nelson to call for the first Earth Day in 1970. Soon after, major federal environmental laws and policies were created, including the Marine Mammal Protection Act, the Endangered Species Act, the National Environmental Protection Act, and the Coastal Zone Management Act. Today, the memory of the spill remains a catalyst for local environmental organizations and elected officials who remain opposed to offshore oil development.

With great irony, the Deepwater Horizon oil catastrophe in the Gulf marked the fortieth anniversary of Earth Day. The long-term consequences of British Petroleum's horrific spill in 2010 remain an issue of litigation and scientific study. Hundreds of thousands of birds were lost. Sea turtles, whales, dolphins, and other marine organisms suffered as well. Many coastal areas associated with the Gulf's petrochemical industry are wastelands. Louisiana is the most polluted state in the United States as a direct result of years of extracting resources and inappropriately disposing of hazardous materials. Currently there are twenty-one Superfund sites in Louisiana. The area is known as "toxic alley." There is no easy fix to the toxic products of our reliance on fossil fuels. Unlike the anti-oil sentiment of Californians, the Gulf's residents remain in favor of oil development. Public perception of the oil industry along the Gulf coast is very different from the opposition to offshore oil held by most coastal Californians.

In the darkness of an oil catastrophe resides the ethical compromise of climate change and our continued overdependence on fossil fuels. Catastrophe is a teacher; the event can represent the beginning of a renewed social capacity for ways of knowing and organizing. In each story of catastrophe, we can find the bridge between the real horror, despair, and savagery of human impacts and the sense of reverence that is needed to persevere and protect. With technology, we can change the character of nature, and as a consequence, become more vulnerable and exposed—inadvertent authors of our own distress.

ORGANIC MACHINES: THE MAKING OF SECONDARY NATURE

If you throw a Dixie cup into the sea, over time organisms will be attracted to the cup, and a simple ecosystem may result. Oil rigs are not different; they attract and foster marine life across the water column. The oil machinery in the sea has a unique ecology that reflects the particular ecological setting the rig is in. The current, eddies, and other physical oceanographic processes also influence the ecology of the rig across time and space. The mega-machinery of offshore oil is not divorced from the ecological setting—it should not be thought of as purely a machine, devoid of life. Offshore oil rigs are examples of "organic machines"—the underwater portion of the offshore structure combines the machinery of offshore oil development with organic material. In this sense, the boundary between what we see as "natural" and as "mechanical" is blurred.

Many environmental groups call for the complete removal of these offshore structures and refer to the rigs as "rubbish," while recreational fishers call for the protection of these structures as "artificial reefs" for fishing. These views reflect the tenuous relationship we have constructed in the marine systems where offshore oil development and other mechanical activities (submerged pipelines, piers, groins, and other structures) take place. In *The Organic Machine* (1995), Richard White describes the relationship between human beings and the hydropower development of the Columbia River. The river also includes several species of endangered and threatened plants and animals. In this sense, the organic machine combines both natural and artificial (mechanical) components. Since the 1930s, White shows, hydropower development has denatured the once wild, free-flowing river into an organic machine. White's notion of the organic machine is a useful metaphor to describe the ambiguous relationship between marine ecology (oceanography, biogeography, and climate-related factors in marine areas) and offshore oil and gas structures. For some, offshore oil and gas activity is a threat to marine ecosystems, and needs to be prevented, and after development the structures should be completely removed. For others, offshore oil and gas activity provides "habitat" that enhances economic values, including fishing activities. There are also ethical issues at stake. Removal of an offshore rig requires explosives. Tons of biomass are lost with complete removal of the underwater portion of the offshore rig.

In particular, the submerged offshore oil structures that exist in the marine system may include a range of invertebrates, shellfish (including mussels and oysters), sponges, bivalves, and other marine life, including fishes. In some cases, the marine life may be *attracted to* the rig, while other species may be *produced at* the rig. This issue of ecological production is a hotly debated topic in marine science. It remains unclear whether the submerged offshore rig produces fish, as habitats like coastal wetlands often serve as nurseries for fishes like halibut and invertebrates like spiny lobster. Scientists do not agree that submerged oil rigs are nursery areas for marine life. In some of the offshore rigs in the marine waters of Southern California juvenile fishes are found in the higher water column, or what scientists refer to as the euphotic zone, the top 120 meters of the marine environment. It remains unclear whether these rock fish are attracted to or produced within the water column of the rig. These fishes may be attracted to the rigs from nearby natural reef areas. The euphotic zone is where photosynthesis takes place, so it is a productive area of the ocean. The question is whether the rig attracts or produces fish. Some argue that fish species are attracted to the rig from nearby natural reefs. Over time, the mussels in the underwater portion of the rig will die off and form large mounds at the bottom of the sea floor. The mussel shells, other dead marine life, and the offshore oil structure can combine to create different types of potential habitat for marine species. In Southern California, larger fish are found near the bottom of some rigs.

In view of of the ambiguous relationship between marine life and its presence on the offshore structures, a political and scientific debate over the decommissioning of offshore oil and gas structures is inescapable (Carr et al. 2004). There is no easy answer or policy-related resolution to the debate over the future of offshore oil rigs. Both complete and partial removal of the structure will have some type of ecological impact on the marine system. It may also have an economic impact. If the upper portions of the underwater structure are removed, those species within that part of the water column will be lost. Leaving the rig in place may also have an ecological impact, especially as the rig and structures deteriorate and the toxic metals and other contaminants become part of the marine system. There is also the risk that the rig will be hit by a marine vessel, causing a blowout and oil spill.

There is no scientific consensus on the ecological importance of offshore oil and gas structures (National Research Council 1996). Accordingly, value-based conflict between scientist, policymakers, and the general public is inevitable given the ecological and political issues and concerns associated with offshore rigs. In this case, scientists cannot tell us whether the marine life associated with an offshore structure is "natural" (Shrader-Frechette and McCoy 1994). The future of these organic machines remains an issue of both political and scientific debate, and includes a negotiating process over the various meanings and discourses of "nature," the role of science, and issues related to technology.

THE CHANGING SOCIO-ECOLOGICAL CONTEXT

Oil is essential to our mechanical age. We have yet to wean ourselves from this fuel despite its serious consequences for the planet and society. The continued use of fossil fuels is part of the unraveling of the biocultural systems of the planet. Our drive to consume oil affirms a violent degradation of the most vulnerable natural qualities of our own genealogy and evolution with the planet. For over a hundred years, the Gulf of Mexico and Southern California marine areas and coastal communities have represented alternative contexts in which the debate over offshore oil is taking place. The natural history, demographics, resource use, and ecology associated with these marine and coastal areas are very different. These differences shape the politics over the decommissioning of offshore oil structures. As we lose many essential services provided by healthy marine ecosystems, there is an increasing call to convert offshore rigs into "artificial reefs" to enhance the production of fish protein.

Ecology and the Gulf

The Gulf of Mexico is a semi-enclosed oceanic basin that is bound by the North American continent and Cuba. It stretches nearly 900 km from north to south and approximately 1,595 km from east to west, with a total surface area of 564,200 km² (Darnell and Defenbaugh 1990). Driven by offshore winds, warm tropical and subtropical Caribbean waters enter through the Yucatan Channel (176 km wide) into the eastern portion of the Gulf. Larvae, juveniles, spores, pelagic fishes, and plant materials are transported into the eastern Gulf and become part of the Gulf Stream.

Biological production is generally high near river outfall areas, which deliver nutrients directly into coastal waters. The Mississippi River accounts for nearly 64 percent of the freshwater input into the northern Gulf, bringing a high annual volume of sediments, nutrients, and anthropogenic pollutants to the delta and the surrounding continental shelf (McGinnis, Fernandez, and Pomeroy 2001). The Texas-Louisiana continental shelf experiences annual hypoxic episodes during the summer months (June to August), when river-borne nutrients cause phytoplankton blooms and the subsequent biological decomposition starves surrounding bottom waters of essential dissolved oxygen. More importantly for offshore oil and gas activities, the delta supplies silt and mud to the broad, flat shelf area where many offshore structures are located.

Although occasional small shoals and rocky ridges occur throughout the Gulf OCS, this broad and vast area is relatively barren, with little habitat diversity. In contrast to the ecology of the marine ecosystems off California, there is very little hard reef substrate in the Gulf.

The Gulf region, Louisiana in particular, is dependent on offshore oil and gas activity for economic stability (Freudenberg and Gramling 1994). Texas and Louisiana initiated offshore oil and gas development in the Gulf region with lease sales in 1920 and 1922, respectively. But these leases were only developed two decades later. On August 14, 1945, the Louisiana Minerals Board held its second lease sale. Kerr-McGee acquired 43,000 acres in shallow water, nearly 43 miles south of Morgan City, Louisiana. In 1947, Kerr-McGee drilled the first commercially successful offshore well in 18 feet of water, approximately 11 miles from shore. Gulf OCS oil and gas activity intensified in the 1940s and 1950s (Wermund 1985). As of September 2012 there were 2,996 platforms in the Gulf of Mexico. Today, well drilling is occurring in over 8,000 feet of water.

Most of the oil and gas structures are set in the soft, sandy bottom of the Gulf, where there exists very little hard substrate for natural reefs. Like natural reefs, studies show that offshore structures attract bacteria, algae, and invertebrate and fish species. (Offshore rigs may also attract nonnative marine species.) Many of these species are economically valuable and commercially and recreationally sought-after. More than one-quarter of the remaining platforms in the Gulf are over twenty-five years old, and may be removed within the next ten years (MMS 2000). This means that nearly a thousand structures may be removed in the next ten years.

Many coastal inhabitants of the Gulf regularly fish coastal and marine waters for recreation, sport, commerce, and subsistence. In Louisiana, over 70 percent of all recreational angling trips occur near the platforms in federal waters (Reggio 1987a, 1987b). For this reason, a number of government and nongovernment organizations began to campaign for changes in public attitudes and existing laws to facilitate the use of petroleum platforms as artificial reefs for fish concentration (Harville 1983). The oil and gas industry was also anxious to cooperate with responsible artificial-reef developers willing and able to accept future responsibility and liability for rigs. A strong coalition of fishery and oil interest groups has emerged, during the long history of oil development in the Gulf OCS region, that supports the oil industry's presence in the region (Freudenberg and Gramling 1994).

Ecology of the Southern California Bight

Unlike in the Gulf, there is abundant rocky reef habitat off Southern California. Over 70 percent of this is around the Channel Islands (Dailey, Reish, and Anderson 1993). Compared to the natural reef areas, the submerged area associated with offshore oil structures represents but a fraction of the hard reef substrate the marine life depends on.

California OCS oil and gas structures are located and embedded in the Southern California Bight (SCB), a marine ecosystem that includes an area between Point Conception (in south-central California) and Punta Banda (south of Ensenada, Baja California, Mexico; Dailey, Reish, and Anderson 1993). This is one of the most studied marine ecosystems on the planet. The ecology of the SCB is a contributing factor in the scientific and political debate over the future of oil rigs. The biomass associated with California OCS oil and gas structures depends on the dynamic relationships and linkages that exist in the SCB (Schroeder and Love 2004). Life around a platform should not be thought of as a "steady state." The particular species present at any given platform depend on the biogeographic setting of the platform and its depth, as well as other factors.

The SCB includes the Santa Barbara (SB) Channel, where the warm water of the Southern California Countercurrent mixes with the cooler water carried by the California Current. Upwelling often occurs where these water masses meet, near the massive headlands of Point Arguello and Point Conception, as well as along much of the California coast, depending on the season. Oceanographic thermal fronts are abundant in the SB Channel; they form as a consequence of upwelling and of current shear between the two primary currents (Harms and Winant 1998).

OCS oil and gas production has occurred in the SCB since the mid-1960s. Most of California OCS oil and gas platforms and structures will need to be decommissioned within the next fifteen years. To date, only seven relatively small structures have been removed from state waters. The most recent project occurred in 1996, when Chevron removed Platforms Hope, Heidi, Hilda, and Hazel. These platforms were in water from 100 to 140 feet deep. One hundred and thirty-four wells were plugged and abandoned on these platforms. In order to remove the rigs to be brought ashore for recycling and disposal, explosives and heavy machinery were used to tear them from their foundations. The biomass that had accumulated around these OCS oil and gas structures was destroyed during the decommissioning activity. The structures that were brought on shore ended up in landfills.

One primary factor influencing the ecology of a rig is the size and amount of biomass associated with the rig. After a decade of fieldwork around several of the rigs, Milton Love, a marine biologist who has been studying rockfish since the early 1960s, showed that 90 to 95 percent of fish near rigs are rockfish. Populations of rockfish off Southern California have dropped to 8 percent of their 1960 populations (Love, Caselle, and Van Buskirk 1998). The significant decline of several rockfish species is a result of overfishing by recreational and commercial fishers. A few of the rigs in the SB Channel include juvenile bocaccio, which is listed as a threatened species, in the upper water column. Rockfish represent $3 billion in sport fishing value and $200–300 million annually in commercial landings.

Offshore oil structures do not serve the same ecological functions as natural reefs (Holbrook et al. 2000). If fish are *attracted* to offshore rigs, then the rig represents merely a "fishing hole" for fishers. This can be a problem because rare species such as the bocaccio rockfish are found at a few of the rigs. If the rig *produces* fishes, then protecting the rig as a type of refugia or marine protected area is warranted.

Fishers are divided over the future of rigs (McGinnis, Fernandez, and Pomeroy 2001). Unlike in the Gulf, Southern California's commercial fishers strongly support the complete removal of the offshore rigs. Trawlers, in particular, have been strongly opposed to any alternative but complete removal. They cite assurances made to them by the off-shore oil industry and the agencies involved that once production ceased, offshore platforms and all associated materials would be completely removed, and their traditional fishing grounds restored. Recreational fishers remain advocates for "reefing" of rigs.

THE POLITICS AND SCIENCE OF DECOMMISSIONING

There are alternatives to complete removal of an offshore rig. The three decommissioning options are:

1. Leaving the rig in place. This option involves steps to insure that oil extraction activity is shut down as well as preparation of the rig to support other uses. The rig is stripped of all equipment directly related to the extraction of oil. Wells are plugged normally; conductors are severed and removed completely to 15 feet below the mudline. All other parts of the rig remain, including potentially much of the above-surface structure. The most biologically important part of the rig is the upper part, and this is likely to be removed in the process of the other two options.

2. Complete removal of the offshore rig from the ocean. The material from the rig is removed from the ocean for multiple destinations (scrapping and reuse or onshore disposal).

3. Partial removal of the oil rig, with disposal of the material either offshore or onshore. Partial removal encompasses several possibilities. All scenarios require wells to be plugged and conductors severed and removed.

With respect to leaving the rig in place, a number of potential other uses have been proposed. Many of these options for reuse of structures are far-fetched. There have been proposals to use the above-water parts of the offshore structures as prisons, hotels, or casinos, or to generate wind energy. Transfer of ownership and liability for the rigs would be required. In addition, the submerged areas could be used as future aquaculture areas for shellfish, oysters, tuna, sea bass, or other marine species of economic value. There have been proposals to develop aquaculture around the submerged offshore oil rigs. The more likely future use is for the current leaseholder to propose development of other leases, but this will require a permit and environmental review of impacts and physical constraints. One incentive the oil industry may have is to keep the ownership of an existing rig after its economic use has ended so that an untapped existing or future lease and oil reserve can be developed that is currently not in production. This would also likely require an extension of the lease granted to the industry by the state and federal governments. In California, many of the rigs share a common pipeline infrastructure for transporting extracted oil. The pipeline is required for transporting oil from the drilling site, rather than relying on vessels or oil tankers. There are also technologies, such as slant drilling, that can enable oil extraction in existing or future lease areas well beyond existing drilling areas.

Organic Machines as Artificial Reefs

A coalition of oil and fishery interest groups, federal and state resource agency personnel, and other individuals, such as congressional representatives, has favored a rigs-to-artificial-reef alternative to complete removal of Gulf OCS oil and gas facilities. For example, in 1979 the Sport Fishing Institute initiated action by urging a resolution to the Secretary of Commerce and the Secretary of the Interior to develop policies, procedures, and guidelines to convert platforms to artificial reefs. The institute also pointed out the importance for fisheries of existing artificial reefs in the Gulf. For the Gulf region, the idea of rigs-to-reefs is less an invention and more a mutation of an old idea. Along the Gulf coast, everything from refrigerators to cement, tires, and junked cars has been used as artificial reefs.

The rigs-to-reefs idea represents the coupling of an already familiar activity of building artificial reefs in the Gulf. The rigs-to-reef policy idea represents a recombination of an old solution (the reliance on artificial reefs to enhance fisheries) to a perceived new problem (the lack of natural habitat and potential economic impacts associated with complete removal of Gulf OCS oil and gas structures). In July 1983 Congressman John Breaux of Louisiana, along with seventeen other members of Congress, introduced H.R. 3474 in an effort to establish a national artificial reef policy. Testimony from federal, state, and local agencies as well as user groups and Gulf conservation organizations supported the establishment of a comprehensive federal artificial reef program that included a rigs-to-reefs option. Some portions of the bill were amended, and it was reintroduced as H.R. 5474. The amendments were approved and passed in April 1984.

The National Fishing Enhancement Act (NFEA) of 1984 (Title II of Public Law 98623) was passed by Congress and signed into law by President Reagan in November 1984. The NFEA defines an artificial reef as "a structure which is constructed or placed . . . for the purpose of enhancing fishery resources and commercial and recreational opportunities." The NFEA states that "properly designed, constructed, and located artificial reefs . . . can enhance the habitat and diversity of fishery resources; enhance recreational and commercial fishing opportunities; increase the production of fishery products in the United States; increase the energy efficiency of recreational and commercial fisheries; and contribute to the United States coastal economies."

The NFEA consolidated several decades of localized and state laws to maximize the potential benefits of artificial reefs as fishery enhancement mechanisms. State governments are responsible for carrying out the general goals of the NFEA in federal and state waters by funding, promoting, and maintaining artificial reefs. The NFEA provides a foundation for the establishment of a national artificial reef program based on the following goals:

- to enhance fishery resources
- to facilitate access for recreation and commercial fishing
- to lessen conflicts between users of marine resources
- to minimize environmental risks
- to follow principles of international law
- to prevent unreasonable obstruction to navigation
- to promote consistency with the National Artificial Reef Plan.

The NFEA directed the National Marine Fisheries Service to develop the National Artificial Reef Plan within one year. The plan was published one year later through the combined efforts of fishermen, divers, scientists, and state and federal resource agencies. The plan serves three purposes:

- to provide guidance to individuals, organizations, and government agencies on technical aspects of artificial reef planning, design, siting, construction, and management for effective artificial reef development

- to serve as a reference for federal and state resource agencies involved in artificial reef permitting
- to ensure that the national standards and objectives established by the NFEA are met.

The plan serves as one guide to developing state artificial reef programs. It includes information on design criteria, permit compliance, management methods, and ideas for increasing artificial reef development. In addition, the plan emphasizes the need for research and monitoring of artificial reef activity.

Gulf State Artificial Reef Programs

Passage of the NFEA was a result of a strong coalition of fishers, members of the oil industry, and politicians who were interested in creating enabling legislation to allow a rigs-to-reefs alternative for the Gulf. Artificial reef programs in the Gulf are carried out by local governments, private parties, and individuals. In 1954, Alabama initiated the first artificial reef program in the Gulf. Natural reefs are virtually nonexistent in Alabama waters. As a result of artificial reef building, the state is now referred to as the "red snapper capital of the world," because Alabama waters provide the highest catch of red snapper in the Gulf. In the Gulf, the primary purpose of the rigs-to-reefs policy is to enhance fisheries (Murray 1994). Most state marine fisheries agencies have assumed the lead in developing artificial reefs through state-developed programs. State natural resources agencies currently direct or coordinate with local agencies in obtaining permits, maintaining liability, financing, constructing, researching, and monitoring marine artificial reefs. Many coastal states have adopted state-specific plans based on the guidance of NFEA.

As of September 2012, there were 2,996 platforms in the Gulf. Approximately 420 (or about 10 percent) of all the platforms had been converted into artificial reefs. By the end of 2013, an additional 359 offshore structures were scheduled for decommissioning, with roughly an equivalent percentage converted into reefs. By the end of 2012, Louisiana had 302 rigs converted into reefs, while Texas had 103, Mississippi 8, Alabama 4, and Florida 3.

There remain a number of issues and concerns associated with the use of rigs as artificial reefs in the Gulf region. In an analysis of interviews with state artificial reef managers, Murray (1994) describes the current status of artificial reef programs in relation to administration, budget, siting, promotion, education, evaluation, future trends, and major concerns. The major issues and concerns that state artificial reef managers raise are as follows.

The liability issue. Since the publication of the National Artificial Reef Plan, which recommended that the U.S. Army Corps of Engineers develop specific permit standards and conditions, the issue of liability remains vague and unclear. The Corps has developed a policy requiring the permit holder of an artificial reef to prove adequate liability coverage. States with reefing programs have assumed the role of the permittee. This has necessitated a close review of the role of the states and localities in implementing the NFEA. As Murray writes, "Because most private fishing associations cannot afford the insurance premium, many states have assumed the role of the permittee. Although many state managers welcomed this as a way of gaining control of artificial reef activities, it has necessitated a closer inspection of each state's liability. The level of concern varied widely, but the general consensus was the clarification is needed from the state's attorney general's office. Most reef managers felt that even this would be vague and subject to interpretation until a case comes before a court" (965).

Scientific uncertainty: production versus aggregation. State artificial reef managers remain concerned about the production-versus-aggregation question. It remains unclear whether rigs attract or produce fishes. As Murray writes, "One troublesome issue is related to the inability of artificial reefs to assist fishery production at all stages of the life cycle" (966). Artificial reef managers are concerned that too much emphasis has been placed on adult fishery enhancement activity and not enough on restoring essential coastal processes, such as estuarine habitats and wetland ecosystems (the "nurseries of the sea").

Limited funding. State artificial reef program funding remains a major concern of administrators and managers. In 1988, the average reported annual budget for reef programs was $139,000, with a range of zero to $400,000 (962). Most funds are generated from either state appropriations or Wallop-Breaux funds, which refers to the 1984 Wallop-Breaux Amendment to the Federal Aid in Sport Fish Restoration Act (16 U.S.C. sec. 777 (1988)).

California's Rigs-to-Reefs Policy

In contrast to the Gulf experience, the rigs-to-reefs option was not accepted by Californians until after twelve years of negotiation (McGinnis, Fernandez, and Pomeroy 2001). While scientific evidence, funded primarily by the U.S. Department of the Interior, shows that some rigs may serve as habitat for marine life, there have been no conclusive findings. This is, in part, the consequence of the scientific uncertainty that is associated with a very dynamic and complex marine system. There remains a paucity of information on the ecology of the rigs of the south coast. Given the anti-oil sentiment of Californians, the rigs-to-reefs alternative has not been accepted by a number of local conservation organizations, including the Environmental Defense Center and other anti-oil groups. Nationwide conservation organizations, such as the Ocean Conservancy, have come out in support of the use of rigs as reefs. There remain a number of technical, scientific, and economic questions regarding the adoption of an alternative to complete removal of existing offshore oil structures.

The first bill to propose rigs-to-reefs for California was SB 2173, introduced by senator Bruce McPherson in February 1998. Entitled "Artificial Reefs," it was cast as a proposal to extend the state's artificial reef program to the Outer Continental Shelf off the California coast. As introduced, the bill cited declines in Southern California's marine species and their adverse effects on the state's recreational and commercial fishing industries as the basis for expansion of the state's artificial reef program, both within state waters and into federal waters. Moreover, it declared artificial reefs' ability to duplicate natural conditions and induce production. Neither of these arguments proved convincing to elected officials; SB 2173 died only four months into the 1998 legislative session.

The rigs-to-reefs idea was revived and reintroduced by senator Dede Alpert of San Diego at the beginning of the next legislative session, in January 1999. SB 241, "Decommissioned Oil Platforms and Production Facilities: California Endowment for Marine Preservation," was very similar to its predecessor in some respects, but also reflected greater attention to issues of technical feasibility, value acceptability, and resistance to budget and political constraints. This bill required that rig to-reef sites become marine reserves, and therefore broadened its appeal beyond that of SB 2173, which had been directed toward the narrower interests of recreational fishing and tourism in southern California. As a marine reserve, a rig-to-reef conversion would contribute to coast-wide efforts to protect and restore the marine environment and fisheries. Recreational fishers supported this bill. (This is a peculiar development given that the recreational fishing industry remains strongly opposed to the state's designation of marine protected areas, as described in chapter 8.) SB 241 highlighted the cost savings to be realized by rig-to-reef conversion, and asserted that those savings should be shared with the citizens of the state, not just recreational users in Southern California. The bill's provision of a funding mechanism based on these cost savings is one element of the proposal directed toward insuring its long-term fiscal viability. As noted above, the bill requires that a portion of the cost savings, to be calculated based on the platform's depth, be split between a newly created Artificial Reef Endowment Fund and the California Department of Fish and Wildlife. Unlike SB 2173, the bill failed to specify that formula, perhaps raising questions as to whether the resulting funds would be sufficient to cover program costs. This second bill passed the state legislature, but was vetoed by Governor Davis because of issues of liability, the ambiguous formula set for cost savings, and the scientific question of whether or not rigs would attract or produce marine life.

In 2001, the Assembly and the Senate both approved Dede Alpert's second bill, SB 1. This third legislative bill proposed to create an artificial reef plan similar to those in the Gulf States. However, the governor vetoed this bill as well, with a note praising the reasonableness of the provisions but claiming that the lack of conclusive scientific evidence made such a proposal premature. Adopting a model akin to the "Gulf experience" was cited as a major concern. But looming on the horizon off the shore of California, twelve of southern California's forty-three active oil rigs will stop producing oil within the next decade, and will have to be decommissioned. The rigs-to-reefs initiative was bound to return in the future.

The three bills showcase the roles diverse values and science play in the politics over decommissioning options. The oil industry that operates the offshore rigs wants to achieve financial savings by avoiding total removal of offshore rigs. This was referred to as the "cost savings" factors introduced in SB 241 and SB 1. However, it remained unclear how much the industry should save and contribute to some type of state fund. It also remained unclear why the industry should save *any* amount of cost associated with an option other than complete removal. Commercial fishers want to be able to fish without the risk of catching their gear on oil structures that remain after partial removal. Local conservation organizations distrust the oil companies, and fear a lack of liability for unanticipated problems if rigs are left in place. Recreational fishers want the opportunity to catch more fish, and believe that rigs add to the existing marine habitat and support rockfish populations.

A fourth bill, AB 2503, was signed in September 2010 by governor Arnold Schwarzenegger. California's rigs-to-reefs policy offers a potential path to partial removal of offshore rigs. The law allows a platform owner or operator to design a partial-removal plan for a platform, and to apply for permission to implement it. The law charges three state

agencies within the California Natural Resources Agency with review-
ing the application: the Department of Fish and Wildlife, the Ocean
Protection Council, and the State Lands Commission. Key aspects of
the policy include:

1. *Cost savings.* The State Lands Commission must determine "cost
   savings resulting from the partial removal of an offshore oil structure compared to full removal of the structure," and the owner or
   operator must pay all this money to the state before approval of
   partial removal. The legislation also provides that between 55 and
   80 percent of the avoided removal cost (as calculated by the State
   Lands Commission, based on information provided by the platform
   operator) would go to the state, depending on what year the state
   authorizes the partial-removal plan. Most of this money will accrue
   to a new fund called the California Endowment for Marine Preservation, dedicated to conservation of marine resources.

2. *Impact assessment.* All partial-removal projects must comply with
   the California Environmental Quality Act, which requires agencies
   to evaluate all potentially significant environmental impacts of a
   proposed project, consider alternatives to the project, and mitigate
   all significant impacts to the extent feasible. The Department of
   Fish and Wildlife must prepare a management plan for after the
   partial removal, which will be reviewed during a number of public
   hearings. Public comments on the proposed plan also need to be
   considered in a final plan for partial removal. The owner or
   operator must provide funds for all the state's activities relating to
   the decommissioning procedure, as well as "sufficient funds for
   overall management of the structure by the department."

3. *Liability and ownership.* The owner or operator must agree to
   indemnify the state against all liability claims, including "active
   negligence," including costs of defending against those claims; the
   indemnification may take the form of "an insurance policy, cash
   settlement, or other mechanism as determined by [the Department
   of Fish and Wildlife]." The owner or operator retains continuing
   liability under any law associated with seepage or release of oil.
   The state must take ownership of any platform in federal waters
   before it may be partially removed.

The federal government estimates that within five to twenty years, all
the oil and gas platforms off the California coast will stop producing oil

and gas in quantities sufficient to be economically viable. Despite this new policy in California, a range of issues and concerns were not fully considered before the passage of the act. Eight of California's platforms reside in water over 400 feet deep, and some sit in over 1,000 feet of water. By contrast, no fixed platform at a depth of more than 400 feet has ever been decommissioned in the Gulf of Mexico (or the North Sea). The depth and mass of most of California's platform jackets make their future removal more complex and costly than for the shallow-water platforms removed so far from the Gulf of Mexico. Removal of the deepest jackets would be a much more complex project than any other removal performed anywhere in the world to date. The removal of large rigs will require the use of barges from the north Atlantic, which will need to travel to the Pacific coast. A substantive amount of greenhouse gas emissions will be produced in the partial or complete removal of large offshore oil structures in California. There is currently no requirement to mitigate greenhouse gas emissions from the decommissioning of offshore rigs.

Studies to support the development of the rigs-to-reefs policy in California did not thoroughly evaluate all the costs and benefits of partial removal, nor did they study the ecological benefits or harms from partial removal of any particular platform (Hecht 2010). The shell mounds on the ocean floor typically contain drilling byproducts such as hydrocarbons and metals, so they will likely require remediation, mitigation, or removal to protect marine life from contamination. There is currently no landfill in California or along the West Coast of the United States that will accept the partially removed structure associated with an offshore rig. The removal of large offshore structures is very costly, and there is currently no recycling plant that will accept this amount of scrap metal in California. Professor Sean Hecht (2010), the co-executive director of the Emmett Institute on Climate Change and the Environment at UCLA School of Law, writes: "Overall, the law is flawed. It puts oil platform operators in the drivers' seat, constrains the discretion of our State agencies to protect the environment, and may subject the State to uncertain future liability. These flaws will make it difficult for the State to develop a rigs-to-reefs program based on sound policy."

ABANDONING THE MYTH OF THE MACHINE

The reliance on artificial reefs (or in this case, rigs as artificial reefs) to enhance fish and other marine life represents an ethical dilemma that science and scientists cannot help resolve. An artificial reef is an organic machine, a type of *secondary nature* produced in this Anthropocene age. Ecology informs us that human beings are part of nature, and so are our machines. The anthropocentric marriage of nature and machine is not a happy one. Ecological process is changed in the mechanization process. The boundaries between perceived nature and the machine are blurred. But the fact is that only healthy ecosystems can sustain society, and our increasing dependence on artificial means to produce protein from the sea is a precarious choice to make. Our machines cannot emulate the spring migration of birds or the return of wild salmon to rivers or creeks any more than they can imitate the complexity and function of marine ecosystems. Moreover, our dependence on the mega-machinery of offshore oil can lead us down the path of catastrophic oil spills, climate change, and loss of health in the marine systems that we are irrevocably connected to.

There are many indigenous mythologies that speak to the plight of a place-based community when the sea is wasted, polluted, and degraded by human activities. One common story is an ancient Inuit myth called "The Mother of the Sea," told in the villages on the north coast (Holtved 1966/67). One version of the myth begins with a young girl who is out hunting with her father. As the two set out by kayak out to sea, the father wants to do away with the girl since she refuses to marry. He throws her overboard. As she hangs on to the kayak, he takes out his knife and cuts her fingers off. She falls into the sea, falling deeper, to the bottom of the ocean. Her fingers have followed her descent, and become animals—walrus, whale, fishes, and seals. She settles on the bottom of the sea as a woman, and finds comfort in a large undersea house, which is full of rooms of sea life. She takes up residence in the house with the other marine animals.

The woman recognizes her love for her community, so she provides the people with animals for hunting. The food from animals begin to melt the ice around the human heart. Unfortunately, the human beings become inattentive to the gifts she has offered them. They throw their trash into the sea, and this trash gets tangled into her hair. Since the woman has no fingers, she cannot remove the trash from her hair, and she calls back the animals that the people had grown accustomed to hunting. The climate around the human heart is cooling.

In the absence of food, the people hold a ceremony. In the ceremony, an *angakkut* (shaman) is chosen to descend into the deep sea to meet the woman. As he gets to the bottom, he moves against the current until he comes to the house of the Mother of the Sea. There he finds a large woman with a face so swollen she cannot see anything. After a fierce fight, he manages to clean her of all the dirt and trash; he combs her hair so that the trash is removed. When he is finished, she tells him that she will repay his friendliness with the animals to hunt once again.

This myth reflects a place-based sensibility and ethic that contributes to a response to the pressures of pollution, the ontology of nature, and the necessity of prey species that can be hunted by human beings. The presence or absence of prey is linked to pollution. The fingerless Mother of the Sea is an overseer of the animals. The people depend entirely on the animals, but the availability of animals to hunt is connected to their behavior; that is, it matters whether people behave properly in their interactions with their wastes, with each other, and with the members of the more-than-human community. The myth symbolizes a deep connection between the ontology of animals and the ethics of people and the places they inhabit. Break a rule, and the animals will make themselves less visible and less available; behave properly, and there will be plenty of them to hunt.

There is an ethic that supports our shared mechanical sensibility that fosters reliance on our organic machines. In some cases, the mechanization of the natural world contributes to the *denaturing* effects of our faith in industrial growth; our dependence on machines reduces our recognition, understanding, and sensibilities that are derived from a more intimate connection and relationship to the life-giving values of ecosystems; simplifies or homogenizes the natural systems we depend on; and reduces the services that healthy ecosystems provide us. We developed technological and scientific instruments to make the exploitation of nature more effective and efficient. But we remain unable to control nature, and cannot manage the industrial catastrophes of large oil spills or so-called "natural disasters." In fact, the common reference to *natural* disaster represents a failure to acknowledge and conceptualize the ethical ramifications of our mechanistic ways. We engineer and transform natural systems with our modern technology. Wetlands are destroyed for oil development, and important nurseries for the sea are also lost. In wetland loss, we have diminished the natural buffers that these wetlands provided against strong storms and hurricanes, which eventually damage coastal towns and cities. There are natural factors that contribute to natural disasters, but they are more often than not consequences of our mechanization and re-engineering of ecosystems.

Offshore oil development moved farther and farther offshore, to the point that it is no longer visible from the shore. The silhouette of the offshore rig recedes well beyond the horizon. Yet, we witness the spilled oil on our beaches, on our feet, in our food, in our hair, and in the air. This is the darker side of the profit and greed that foster oil exploration and development. It is hard to "green" the oil we depend on, no matter how proficient our engineers become. Our response to oil spills is yet another problem. We negotiate the spill by sterilizing coastal and marine areas blemished by oil by using detergents. These detergents may make the spill less visible, but there are also ecological impacts from the toxins in these detergents. The concentration of detergents and other chemicals used to respond to oil catastrophes can lead to other environmental nightmares (Hazen et al. 2011).

The reliance on offshore oil structures as artificial reefs is our Frankenstein's monster, an ironic twist of fate of our own making that is driven by our myopic dependence on oil, and the continued ecological destruction of marine systems brought on by its use—now reflected in burgeoning climate disturbance and supported by our further exploitation of the high seas. As we have denatured marine ecosystems, we have grown increasingly reliant on organic machines to provide us with a few of the services the *once* healthy marine systems provided—in this case, both California and Gulf states are increasingly interested in the establishment of artificial reefs to replenish lost fish populations. This technocentric dependence on artificial reefs is a Faustian bargain.

To build an organic machine, modernity continues to transform unique marine places into mechanized artificial spaces that are poor representations of naturally occurring reefs. The organic machine is linked (as for temporary adaptation to a hostile space environment) to one or more ecological processes upon which some of our vital physiological and biological functions depend. The making of the organic machine mirrors the mechanization, capitalization, and objectification of nature into a machine. It is becoming increasingly difficult to distinguish the artificial from the natural. The organic machine stands as the simplified "double" for nature.

We need to think more critically about the myriad impacts of the mechanization of place. An ecologically literate community understands that marine systems are a source of life; they cannot be replicated or replaced. This is the message of the Inuit myth of the Mother of the Sea. Every living thing on the planet depends on a healthy oceanic system, which needs to be something more than a mere mechanical artifice that resembles some of the basic features of naturally complexity of ecosystems. The reliance on an artificial reef, accordingly, will not suffice in the long term. It is an oversimplified replica—even if that replica produces fish for our consumption. Maritime peoples recognize that human life and sustenance depend on a wild, untamable ocean. The ocean is not a machine but a source of cultural and ecological revitalization. Sharing in the ecology of the maritime is the foundation of a more sustainable coastal community and place. If we recognize that we are part of these living systems, then we need to address and respond to the increasing reach of our mechanical sensibility and its consequences.

CHAPTER 6

The Politics of Civic Science

Marine Life Protection in California

El hecho es que hasta cuando estoy dormido
de algún modo magnético circulo
en la universidad del oleaje.

The fact is that until I fall asleep,
In some magnetic way I move in
the universe of waves

—Pablo Neruda, *El Mar*

I grew up along a coastal province and inherited a fish story. Early on Saturday mornings, my father would take me down to the dory fishers at Newport Beach, in Southern California. Small-scale or artisanal fishers sold their local catch to nearby residents. The dory fleet was made up of local commercial fishers who took to the sea in the early light of the morning in small boats that would be launched from the sand with the waves pounding. Under a canopy or tarp, the local fishers would sell their morning catch. We would purchase a fish after a long conversation. We discussed where the fish was caught, the bait used, the tackle employed. I listened; it was a lesson from the sea.

My father told me a story of a great black sea bass. When a large fish took my grandfather's bait, my father wrapped his arms around his father's waist. Grabbing hold of my grandfather, my father was his anchor, holding him down against the pull of the great fish, as my grandfather's fishing pole bent over the boat's rail. But the big eyes of the sea bass did not appear. The fish could not be landed.

There are different stories to tell today. Marine systems are in serious need of protection, and scientists continue to encourage the adoption of marine protected areas (MPAs) across the world's ocean. For some fishers, a designated MPA robs them of their fishing ground. But a good hunter knows when to stop hunting. When the big fish are gone, a good hunter knows that those few big fish that remain are worth protecting, no less than the stories of our grandfathers are worth telling. A fish is not only a fishery. A big fish embodies the life of a fisher's story and knowledge, and plays a fundamental role in the marine ecosystem. Fish are both predators and prey. Without these great fish, our stories along the waterfront or marina are culturally less meaningful. We can't be left with mere memories or stories.

In January 1995, machete-wielding locals protested the creation of the Galapagos Marine Resources Reserve and blocked the entrance to the Darwin Station in Puerto Ayora, the headquarters of the national park. The protesters held several workers of Ecuador's National Park Service captive for four days, and harassed marine scientists. Later, in 1997, twenty masked men on the island of Isabela attacked national park personnel, local officials, and marine scientists of the area. This protest by angry fishers reflects a tension between the need to protect marine life and the values of marine use.

With increasing globalization of commercial fishing efforts, most of the dory and artisanal fishing activities that once supported local economies and maritime cultures have been lost. As the stories are lost, parts of the language and knowledge of local maritime peoples are also set adrift, soon to be forgotten. This chapter describes the politics of setting aside marine areas as no-take reserves off the shore of California. First, I describe evidence of marine ecosystem decline that is increasingly being reported by scientists, and the role of commercial fishing as one primary factor contributing to this decline. I review the role of science and scientists in two planning processes that led to the designation of MPAs in California. Scientists played a role in the politically charged planning effort, but so did members of the powerful commercial and recreational fishing industries. To protect marine life we need to up-scale MPA designation and down-scale the level of economic use of marine life to support place-based marine governance and marine resource use.

THE SLOW VIOLENCE OF MARINE ECOSYSTEM DECAY

The level of coastal marine ecosystem loss is metaphorically akin to a type of slow violence, insofar as society has not responded to the social, economic, and ecological factors that are influencing the health of the ocean. There is no simple response or institutional resolution to the dramatic changes that human beings are inflicting on the world's ocean. Akin to the images of the impacts of an offshore oil spill, like the catastrophe in the Gulf, many impacts on the sea are immediately perceived and recognized. There is an "immediate violence" associated with commercial fishing activity that should be recognized. As of 2014, drift gill nets for shark and swordfish are still used off California even though they are prohibited in Oregon and Washington. Gill nets are vertical walls of netting that are submerged and float freely in the current, catching marine life in an indiscriminate way. Nineteen permits have been granted to fishers by California fishery managers for swordfish and shark. The total commercial value of swordfish landed in 2013 was $2 million. A vast majority of swordfish consumed in the United States is imported. For every five swordfish landed by those who hold permits to use drift gill nets, one marine mammal is also likely to be killed. In May 2014, images of dead whales, dolphins, turtles, birds, and other marine fishes caught as by-catch increased awareness of the impacts of this commercial fishing off California. Oceanus, a conservation organization, filed a federal freedom of information request for the images of by-catch from this fishery taken by observers from the National Oceanic and Atmospheric Administration. Oceanus used the images to garner support from a few California legislators to introduce a bill that would ban the use of drift gill nets for the swordfish and shark fishery. But the bill never made it to the full legislature. Because of the powerful commercial fishing interests the bill was killed in subcommittee and never saw the light of day.

Drift gill nets are used off every continent. In some regions they are used by artisanal fishers to support local communities, while in other places they are used by large-scale commercial enterprises, such as the shark and swordfish fishery. Artisanal fishers are often displaced by larger-scale commercial fishing operations, especially in less developed small island countries of the South Pacific. The fact is that there is some type of by-catch in all commercial fishing activity. There are a range of different gear types, such as the long line, that kill species during the fishing activity. Sea birds, like the sooty shearwater, are caught in the purse and drum seines that dip and scoop California's largest commercial fishery, market squid. Ironically, the calamari you eat at a local restaurant in California was probably caught offshore, sent overseas to Europe or Asia, packaged, and then sent back to the state's consumers.

Imported fish also have a carbon footprint that should be considered when evaluating whether the seafood is "sustainable." Another issue to consider when you choose to eat seafood is that many of the species that are considered by-catch have special protection under many state and federal laws, including the U.S. Marine Mammal Protection Act and the U.S. Endangered Species Act. Marine mammals and birds have a fundamental role in the maintenance of marine ecosystem function and complexity, and these species are rarely considered in fishery plans or by consumers of seafood. Every pound of protein from the sea includes an array of species that were also caught. So your plate of seafood includes not only a swordfish steak but also figuratively speaking part of a dolphin, bird, whale, or turtle. In the shrimp fishing industry by-catch is referred to as "trawl trash." Snow shovels are used to scoop up and throw overboard the dead by-catch that is indiscriminately landed on board. For every pound of live shrimp harvested from the sea, roughly 99 pounds of once living organisms are thrown overboard. The consumer of seafood should be wary of so-called sustainably harvested shrimp or other wild fish species. Few fish species should be considered sustainably caught, especially those that are exported across the world.

The global trade in marine life as commodities contributes to the slow violence that we inflict on marine ecosystem health. We have long assumed that the ocean is a great endless frontier of resources to be used at our discretion. This perception masks the nature of the threats, pressures, and impacts human beings have on the sea. Rob Nixon, a professor of English at the University of Wisconsin—Madison, has a unique way of characterizing the type of perceptual challenge. In *Slow Violence and the Environmentalism of the Poor* (2011b), he describes the slow violence inflicted by globalizing economic forces on nature and culture. In an essay for the *Chronicle of Higher Education*, Nixon (2011a) writes:

> We are accustomed to conceiving violence as immediate and explosive, erupting into instant, concentrated visibility. But we need to revisit our assumptions and consider the relative invisibility of slow violence. I mean a violence that is neither spectacular nor instantaneous but instead incremental, whose calamitous repercussions are postponed for years or decades or centuries. I want, then, to complicate conventional perceptions of violence as a highly visible act that is newsworthy because it is focused around an event, bounded by time, and aimed at a specific body or bodies. *Emphasizing the temporal dispersion of slow violence can change the way we perceive and respond to a variety of social crises,* like domestic abuse or post-traumatic stress, but it is particularly pertinent to the strategic challenges of environmental calamities [my emphasis].

While some impacts are more visible than others, such as images of by-catch from drift gill net fishing for swordfish or the sight of birds on the beach smothered in oil goo, there are less visible impacts from our overuse of marine life. The impacts on the more hidden aspects of marine ecosystems should be a public concern. We know very little about the ecological structure and complexity of large-scale marine systems, such as the links between currents, climate, and biology. This is particularly the case for the benthos or deep sea floor. We know more about the moon than we know about benthic areas. We also do not know the impacts of commercial fishing on food webs, and this issue is rarely considered in governmental attempts to regulate fishing.

There are species that are less visible and more difficult to identify and study. Over the past century, the abundance of phytoplankton, for instance, has declined to the point where major restructuring of the world's marine ecosystems is occurring (Boyce, Lewis, and Worm 2010). There are no special policies or plans to protect phytoplankton, even though plankton is the primary source of productivity of marine systems. Phytoplankton biomass is a critical indicator of the general health of marine ecosystems, and a major producer of oxygen that all life on earth depends on. The health of the planet is irrevocably connected to the production of oxygen by phytoplankton. Without phytoplankton in the sea, you would lose every other breath you take of oxygen.

GLOBALIZATION OF PROTEIN FROM THE SEA

Large-scale commercial fishing efforts are currently overcapitalized and have historically been heavily subsidized by government. Few of us who eat seafood know where the fish was caught, how it was caught, and the impacts fishing has on marine systems and cultures.

We import most of the fish we eat and export most of the wild fish landed in the marine waters of this country. Paul Greenberg (2014) refers to this as the "great fish swap": we are "low grading our seafood supply" to the detriment of our maritime communities and marine systems. "With seafood," Greenberg writes, "we are talking about the destruction and outsources of the very ecology that underpins the health of our coasts and our bodies. . . . In becoming Asia's premier market for shrimp, the United States has effectively unhitched itself from its own seafood supplies and hollowed out its ability and rationale to protect its own marine resources" (3, 5). Unless you purchase a fish from the fisher directly, you'll find that most fish consumed is hardly local or sustainable. We are becoming more vulnerable to the insecurities brought on by the export of our essential protein from the sea, and more reliant on fish of low quality that is imported.

One consequence of the global trade of commercially landed fishes is that unique and sensitive marine ecosystems, such as the Ross Sea of the Antarctic, are showing significant signs of decline (Ainley 2010). One factor contributing to the ecological decline of the Ross Sea is the over-fishing of the Antarctic toothfish (*Dissostichus mawsoni*), which is fished during the austral summer above the Antarctic Shelf (by countries such as New Zealand) and then exported and traded in international markets (e.g., to the U.S. and northern European markets). A majority of the world's fisheries surpassed sustainability in 1988 (Watson and Pauly 2001). Moreover, the basis of the marine food chain has been fished out by the substantive removal of large marine predators from most of the world's oceans (Jackson 2001). Commercial fishers are fishing down the food chains of the world's oceans—over time they have shifted to prey species such as sardine, squid, and mackerel. The shift to prey species is based on increasing global demand for marine fish protein in Asian markets (Pauley 1998). For instance, new markets for California fisheries such as market squid have emerged since 1980 because of new market demands in Asia (CDFG 2001). Indeed, the history of marine life exploitation is one that leads to biological extinction of the species and/or the economic collapse of the fished species.

In a comprehensive history of California commercial fishing activities, Arthur McEvoy (1986, 6) describes the development of industrial-scale commercial fishing in the state as one that has "followed a repetitive pattern of boom and bust, one typical of fisheries the world over." There are currently over 285 species fished and landed commercially (and recreationally) along the California coast (California Fisheries Fund, 2008). California market squid (*Loligo opalescens*) ranks as the state's largest commercial fish landed by volume. Among U.S. exports of commercial fisheries in 1999, squid ranked sixth by volume and sixteenth in value, higher than any other California commercial fish species (CDFG, 2001). A vast majority of the squid is exported to China and European markets. In 2005, the Management Plan for squid was set by the California Department of Fish and Game (CDFG 2005) at a fixed seasonal catch limit of 107,048 metric tons, which is the highest historical landing for the fished species (it occurred in 1999). It is difficult to fathom the amount of squid caught. One hundred and seven thousand metric tons is over two hundred million pounds. That is a lot of food for thought and for marine species. One way to imagine this amount of biomass was proposed by a colleague and mathematician, who estimates that 120,000 metric tons could fill nine Los Angeles Memorial Coliseums to the brim with dead squid.

A vast majority of squid is caught within designated national marine sanctuaries, including the Channel Islands National Marine Sanctuary and the Monterey Bay National Marine Sanctuary. The global scale of commercial fishing for squid off California reflects a particular tragedy of the oceanic commons: if we wish to protect and sustain the multiple ecological values of marine life we need to protect the quality of marine areas and habitats that species depend on while addressing the scale of economic trade of species. Missing from the Market Squid Management Plan is an ecological understanding of the role the species plays in the ecosystem—the species is important to millions of fishes, birds, and mammals (Zeidberg et al. 2006). When commercial fishers mine for market squid in marine *sanctuaries*, the ecosystems of these areas are impacted. The market squid is a principal forage item for a minimum of nineteen species of fishes, thirteen species of birds, and six species of mammals. Squid is considered a keystone species, yet also considered a commodity traded overseas despite its essential value in the system.

A vast majority of the commercially valuable fishes of the world's ocean are traded in global markets, and are severely depleted (Worm et al. 2009). Overfishing is considered a primary factor contributing to the disruption and degradation of coastal marine ecosystems. Jeremy Jackson and colleagues (2001, 629) write, "Overfishing and ecological extinction predate and precondition modern ecological investigations and the collapse of marine ecosystems in recent times, raising the possibility that many more marine ecosystems may be vulnerable to collapse in the near future." There remains an urgent need to protect large areas of marine ecosystems from the impact of fishing and other extractive activities, such as mining and offshore oil development. In an article published in the journal *Oceanography,* Jane Lubchenco and Laura Petes (2010, 3) warn: "Many ocean ecosystems appear to be at a critical juncture. Like other complex, nonlinear systems, ocean ecosystems are often characterized by thresholds or tipping points, where a little more change in a stressor can result in a sudden and precipitous loss of ecological functionality."

Less than 1 percent of the world's ocean is protected in some type of marine reserve. Few marine reserves are successful, and only one in five that have been studied protect marine ecosystems (Halpern 2014). Few reserves protect pelagic species, such as tuna, squid, and other migrating species. Migrating species are rarely protected by marine reserves because of their habitat needs and foraging characteristics. A designated whale "sanctuary" off a particular coast protects part of the marine area that the migrating whale depends on for survival and reproduction. But as the whale moves out of the reserve, it can be caught in fish nets, hit by a large container ship, suffer the consequences of pollution, or be killed by a Japanese whaling vessel. There are few true sanctuaries for marine life across the world's ocean. Some MPAs protect species that do not migrate large distances. Most MPAs can be thought of as paper tigers, insofar as they do not protect the full range of species across the food web, nor do they protect enough high-quality marine habitats to preserve marine ecosystems. For instance, the thirteen areas that are part of the U.S. National Marine Sanctuaries Program regulate a number of marine activities within their administrative boundaries (Farady 2006). A vast majority of marine area within designated sanctuaries allows fishing, and cannot protect species from the impacts of climate change or prevent the threats from poor water quality of marine habitats. Large-scale fishing of marine sanctuaries may not be consistent with the program's priority goal to protect marine ecosystems while allowing marine resource use that is *compatible* with this goal.

MARINE LIFE PROTECTION IN CALIFORNIA

In 1999, a complicated system of MPAs off California included 0.06 percent of the total marine area within state waters. Based on the state's complicated range of MPA designations and the inadequate level of marine protection, the California Legislature approved and the governor signed the Marine Life Protection Action (MLPA). Related state legislation includes the Marine Life Management Act of 1998, Marine Managed Areas Improvement Act of 2000, and California Ocean Protection Act of 2004. The MLPA emphasizes the role of MPAs as a tool in coastal marine ecosystem-based planning. The MLPA planning process included professional facilitators, social and physical scientists with expertise in diverse disciplines, stakeholders, consultants and other advisors, and resources agency personnel. The cost of implementing the MLPA, including the planning processes, has been in the tens of millions of dollars (CDFG 2008b). This section includes a description of the two phases of the implementation of the MLPA.

Phase 1 of the MLPA

In 1999, three advisory groups were established by the Channel Islands National Marine Sanctuary (CINMS) and the CDFG to consider the designation of MPAs for the marine area six nautical miles offshore of five northern Channel Islands in Southern California—the marine areas around Santa Barbara, Anacapa, Santa Cruz, Santa Rosa, and San Miguel Islands (McGinnis 2006, 2012a). The CINMS is one of thirteen sanctuaries managed by the National Marine Sanctuaries Program. The MPA planning effort for the CINMS included a Marine Reserves Working Group (MRWG), a Science Advisory Panel, and a Socioeconomic Advisory Panel (U.S. Department of Commerce, 2007). The MRWG included seventeen members, who were intended to represent a wide diversity of interests and values within the region. They included representatives from state and federal resource agencies, user groups (e.g., commercial and recreational fishers), and local and national environmental organizations, and an academic. The MRWG met from July 1999 to May 2001. The group represented the first collaborative effort to develop and establish no-take MPAs in California, and included two paid professional mediators. The funds for this consensus-based planning effort came from the state and federal resource agencies.

In September 1999, five goals (including the two goals of protecting marine biodiversity and sustaining fisheries) were agreed to by all members of the MRWG (U.S. Department of Commerce 2007). The sixteen-member Science Advisory Panel developed recommendations for MPA designs based on the MRWG goals. In November 1999, the science panel recommended to the MRWG that a network of no-take marine reserves representing 30 to 50 percent of the entire CINMS could satisfy the MRWG goals. Following a comprehensive evaluation of the available scientific information on MPA design, the panel based their recommendations on the following major findings:

- Larger reserves (from 30 to 70 percent of habitat) can protect more habitat and populations of species while providing a buffer against losses from environmental fluctuation or other natural factors.
- No-take marine reserves can enhance species diversity, biomass, abundance, and size of marine animals.
- Case studies of no-take marine reserves show positive spillover effects from reserves into fishing areas that exist outside of the reserve area.

· Reserves that are designed to protect ecosystem biodiversity can also protect commercial and sports fisheries.

The science panel produced forty maps that represented a range of alternative MPA network alternatives covering 30 to 50 percent of the CINMS (U.S. Department of Commerce 2007). The maps included a visual representation or layer that identified the estimated socioeconomic costs associated with MPA alternatives. These cost estimates were based on an analysis of recreational and commercial fishing information, including ethnographic data, that was analyzed by economists at the federal government's National Ocean Service and other social scientists who were members of the socioeconomic panel. The "characteristic" scale associated with each map reflected an estimate of the level of protected representative habitat and an estimate of the related economic cost to the fishing industries for each reserve scenario. There was also an assessment of the economic benefits of the MPA options.

In their deliberation, the members of the science panel reached consensus on a number of major points. MPAs are both a fishery management tool and a biodiversity conservation strategy. The 30-percent recommendation could protect up to 70 percent of the biodiversity within the CINMS, while a 50-percent reserve design would capture roughly 85 percent. Less than 50 percent would not protect birds or mammals; but a reserve design greater than 50 percent would have detrimental impacts on the goal of sustaining fisheries. In addition, the panel noted that the size of reserves should be based on an "insurance" factor to strengthen ecological resilience to major disturbance events and potential human catastrophes such as an oil spill (Allison et al. 2003). With respect to the insurance factor, the panel recommended that a reserve design should include a multiplier (i.e., 120 to 180 percent of the reserve spatial design) in case of catastrophic events, such as an oil spill or a major warming event that can destroy habitats. This insurance factor was described by the panel as essential to reserve design given the ocean-climate variability associated with the marine area. Given environmental fluctuation, some areas within a potential reserve scenario would not protect species because of ecological disturbance to the habitat in the MPA.

The Scale and the Scope of Conflict

Conflict between the members of the MRWG over the scientific recommendations was intense, and it received media attention throughout the state. One institutional challenge was that the formal designation document of the CINMS did not provide a clear mandate to develop MPAs or any other tool to protect coastal marine life. Resource managers and recreational and commercial fishers remained wary of the scientific recommendations. The insurance factor was dropped by most of the members of the MRWG. From September 2000 to February 2001, the MRWG members debated the science and could not reach agreement on the size or location of MPAs.

Even after months of public debate and informal negotiation, the members of the MRWG failed to reach consensus on the scientific recommendations. Local conservation organizations supported the science panel's recommendations during this period, including important organizations such as Santa Barbara's Environmental Defense Center. The members of the MRWG also participated in months of mapping exercises that took place in April and May of 2001 to consider reserve alternatives; the maps evaluated did not protect a minimum set-aside of 30 percent, and there was no consensus among MRWG members on a particular map or reserve design. Only one member of the MRWG supported the science panel's recommendations.

In May 2001 the MRWG was disbanded after failing to reach agreement on a map(s) or the scientific recommendations made by the science panel. The members of the MRWG representing commercial fishers did not support the use of MPAs as a fishery management tool, and did not support an expanded role of the CINMS to manage fishers in the sanctuary (Marine Reserves Working Group 2001). Recreational fishers in the MRWG did not support the use of MPAs, and were in favor of maintaining access to fishing areas located in the eastern waters of the CINMS, including Anacapa and Santa Cruz Islands, which are close to the Ventura Harbor.

Based on the mandate of the MLPA, California Department of Fish and Game staff worked with federal resource agencies to finalize a recommendation for a preferred alternative MPA network designation for state waters. In August 2001, the department recommended to the California Fish and Game Commission the formation of a network of MPAs in the marine areas in state waters (0 to 3 nautical miles) that exist within the CINMS. The state assumed that the federal government would complete the designation process for the federal waters (3–6 nautical miles) in a separate environmental review process. The recommendation represented the culmination of more than three years of deliberation among a variety of groups, and well over ten thousand comment letters were submitted during the state's California Environmental Quality Act review process.

In October 2002 the CDFG Commission adopted regulations to create MPAs within the near-shore waters of the CINMS. The MPA design emphasized the importance of establishing no-take marine reserves. The National Oceanic and Atmospheric Administration expanded the MPA network into the sanctuary's deeper waters (3–6 nautical miles) in 2006 and 2007. The entire MPA network consists of eleven marine reserves where all take and harvest is prohibited, and two marine conservation areas that allow limited take of lobster and pelagic fish. The MPA network encompasses 241 square nautical miles (318 square miles), making it the largest such network off the continental United States.

The general assessment of this first phase of the MLPA process is that stakeholders failed to reach consensus because of the lack of institutional resources and support, and that significant changes to future planning processes were needed that moved beyond the goal of consensus between stakeholders (Osmond et al. 2010; Gleason et al. 2010). One view is that the MRWG process was halted because of a lack of resources, for example staffing, funding, and technical tools (Gleason et al. 2010). An additional assessment is that the role of the scientists in the formal collaborative planning effort needed to be weakened (Helvey 2004). Science or scientists, it is argued, should not drive the planning process. Because MPAs are controversial tools and user groups are often unsupportive of their designation, one view is that the planning effort should be based on a stakeholder-driven process and not a *strict* interpretation of the MLPA mandate (Gleason et al. 2010).

Another view is that the MRWG members failed to reach consensus because of their value-based differences, and the intergovernmental conflict over the scale of protection recommended by the science panel. The recommended scale of protection led to an expanding scope of conflict between diverse interests and value orientations (McGinnis 2006, 2012a). The MPA debate highlights the problem of informal and, at times, formal (i.e., government initiated and supported) modes of the privatization of marine areas or ocean commons by vested interests (McEvoy 1986). Since the designation of the CINMS in 1980, the maritime commons of the sanctuary was informally "privatized" by fishing interests (Turnipseed et al. 2009). The MRWG process threatened the privileged role that commercial and recreational interests play in state and federal ocean resource management and planning because it opened up the decision-making process to a broader and more diverse public.

Up-scaling marine biodiversity protection represented a threat to the scale of resource use in the CINMS. Until 1999, the CINMS represented merely a paper tiger insofar as very little of the marine area (a small area off Anacapa Island) was protected from some type of resource use or extraction. The economic costs associated with protecting 30 to 50 percent of the entire CINMS were found insignificant to the region's economy by economists involved in the process (U.S. Department of Commerce 2007). There would be potential impacts on particular fishers, because a fisher would lose a marine area to a MPA closure.

Gary Davis, who at the time was a senior ocean scientist at Channel Islands National Park, was a member of the MRWG. After the process, Davis (2005) maintained that this planning process exposed the conflict between values and ethics held by diverse stakeholders, which eventually led to a major compromise on the recommendations by the science panel. There is empirical support for this claim. In an analysis of the data collected in 2002 from surveys of stakeholders involved in this first planning effort, a study found at least two "advocacy coalitions"—an anti-MPA advocacy coalition and a pro-MPA advocacy coalition (Weible and Sabatier 2005). The anti-MPA coalition (e.g., recreational fishers) remained opposed to the designation of marine reserves. The MRWG debate also challenged the relationships between government agencies responsible for marine stewardship. Staff in the National Marine Fisheries Service (NMFS), CDFG, and CINMS were engaged in conflict over what role, if any, the federal government should play in the designation of MPAs in federal or state waters; the utility of MPAs as a fishery management tool; the role of the National Marine Sanctuaries Program in fisheries management; and what resource agency or jurisdiction should take the lead in developing MPAs under the National Oceanic and Atmospheric Administration.

Mark Helvey, who was a member of the MRWG representing the NMFS, published an article in the journal *Coastal Management* that represented his position on the MRWG process. Helvey's 2004 essay surprised many who were involved in the planning effort. Currently the assistant regional administrator of the NMFS's Southwest Region, Helvey maintains that there is a need for stakeholders to recognize irreconcilable impasses early in the MPA planning effort and to seek solutions to "maneuver" around conflicts. One could conclude from Helvey's article that when science is not acceptable to a particular interest or user group (e.g., commercial or recreational interest) some form of political maneuver (e.g. lobbying or litigation) is needed.

Maneuvering around the science, in effect, can be interpreted as an attempt to reject *unfavorable* information that threatens one's economic interest despite the clear goals of the MLPA. Helvey's recommendation seems to miss the intent of the MLPA, the goals of collaborative decision-making, and the ground rules that each member of the MRWG agreed to. Helvey's view also challenges the importance of the use of "sound scientific guidelines" as set forth in the MLPA.

Following sound scientific guidelines may be "in the eye of the beholder." In the MRWG process, professional facilitators emphasized the place of values and diverse stakeholder interests in the interpretation of scientific information or, in this case, the recommendations that were made by the members of the science panel. The science panel's recommendations were information that would unfavorably impact the economic return of sport and commercial fishing. Formal facilitation and collaborative decision-making strategies are useful, but these tools do not guarantee that scientific information will be received or acted on by all participants, stakeholders, and interests. In the case of the MRWG deliberation, it is not surprising that the commercial and recreational fishing interests and stakeholders in the process rejected the recommendations made by the science panel. But it is surprising that all federal and state agencies (and the MRWG member from the Ocean Conservancy) failed to accept the recommendations made by the scientists.

Economic Impacts from MPA Designation

Since the MPA network was designated by state and federal governments, there have been relatively minor impacts to commercial and recreational fishing interests (CDFG 2008c). Prior to the establishment of MPAs in 2003, economists predicted that commercial fisheries in the CINMS would decline in economic value by up to 17 percent. Five years after MPA establishment, changes in seven fisheries were analyzed. Three fisheries had declined, including two that decreased more than predicted, and four fisheries had increased, instead of declining as predicted. However, many factors beyond the MPAs played a role in these changes, including other regulations, environmental changes, and market forces. Some commercially fished species, such as rock crab, spiny lobster, market squid, and red urchin, have grown in value, while others (sea cucumber, California sheephead, and rockfish) have declined (CDFG 2008c). Many of these changes occurred throughout Southern California, suggesting causes other than the designation of MPAs. This is especially the case for rockfishes, given that major closures were made by the CDFG before the MRWG process. Detailed studies of the lobster fishery suggest that some changes in numbers of fishers and catch may be linked to the MPAs. The number of party boat trips for recreational fishing has remained fairly constant since the MPAs were established.

Phase 2 of the MLPA

In 2004 a renewed effort to implement the MLPA with a new regional approach to collaborative planning began (CDFG 2008a). The MLPA Initiative established a Blue Ribbon Task Force (task force), together with a Master Plan Science Advisory Team (science team) and a Regional Stakeholder Group, to oversee the completion of several objectives (Saarman and Carr 2013). The first of these objectives was to develop a Master Plan Framework, which would include guidance, based on the requirements of the MLPA, for the development of alternative proposals of MPAs in state waters. The plan also identified specific study regions, with a goal of completing the statewide planning process by late 2011.

The scholarly literature describes this second phase of the MLPA as a successful planning effort. Dozens of peer-reviewed journal articles have been written on this second phase by those who were formally involved in the state's planning effort, for example consultants or advisers, resource agency personnel, and participants in the science team (Saarman and Carr 2013). A number of articles describe the general success of the state's collaborative process, and the role of scientists and stakeholders. This literature, in general, fails to consider the outcomes from phase 1 of the MPA designation process for the CINMS and phase 2 for the south coast region, which focused on the coastal mainland (not the islands). I turn to a comparison of these two planning processes below.

A COMPARISON OF THE PHASES OF MLPA PLANNING

A comparison of phases 1 and 2 shows that the management preference has tended to emphasize less protection for marine life and a more concerted effort to address fisheries issues during phase 2, by adopting a "network" approach to reserve design and dropping the goal of "consensus-based" planning in regional decision-making. The members of the science team focused on network design, rather than on some level of minimum set-aside as in phase 1. Network design emphasized the potential benefits of small reserves to fished species, and general habitat connectivity across both no-take marine reserves and conservation areas that allowed some type of fishing activity. Neither phase 1 nor phase 2 processes considered the large-scale economics of commercial fishing off California. Table 6.1 depicts the level of MPA set-aside for the completed processes, and compares the outcomes of the planning efforts with the first phase.

Compared to phase 1, the phase 2 designation outcomes include:

- an emphasis on "fishery-based science" and network reserve design rather than larger-scale biodiversity conservation
- a shift from the designation of no-take MPAs to conservation areas that allow some type of fishing or use
- a reduction in the physical scale and level of protection provided by MPAs
- an emphasis on garnering "broad-based agreement" among a high number of stakeholders, rather than consensus
- a change in the scientific criteria that supports the role of small reserves based on MPA network connectivity rather than the size, quantity or scale of representative habitats protected in the MPA network.

What explains the difference in the level of protection, given that many of the same scientists were involved in both phases of the MLPA planning process? Table 6.1 shows the dramatic decline in the designation of no-take MPAs by the state, as indicated by the 21-percent no-take MPA network created for the CINMS marine areas (established during the first phase) and 4.9 percent for the south coast study region, as adopted in December 2010 by the CDFG Commission (phase 2).

One explanation is that powerful fishery interests carried the day, and influenced the process of the state's designation of MPAs. The south coast planning process concluded after two years of negotiation among sixty-four stakeholders. The CDFG Commission adopted regulations to set aside an additional 187 square miles, or 8 percent of the state waters of the south coast (the existing MPAs around the northern Channel Islands encompass 168 square miles and 7 percent of state waters in the study region). Even if the total percentage of MPAs in the CINMS is combined with the state reserves off the coastal mainland of the south coast, one recognizes a major shift in the type of reserve designated and the scale or level of protection provided to marine life.

TABLE 6.1 STATE AND FEDERAL MARINE PROTECTED AREAS OFF CALIFORNIA

Phase and region	No. of stakeholders	Percentage of no-take marine reserves	Type of planning process and procedural outcome	Support by the scientific advisory group	Percentage of total reserves*
Phase 1					
Northern Channel Islands (established in state waters in 2002, and federal July 29, 2007)	17	21	Consensus-based Agreement on problem statement and goals of MPAs No consensus on science or reserve design	No—scientific advisory group supported 30–50% MPA network	22
Phase 2					
Central coast (took effect September 21, 2007)	56	7.5	Broad-based agreement among regional stakeholder group	Yes	18
North-central (took effect May 1, 2010)	45	11	Broad-based agreement regional stakeholder group	Yes	20.1
South coast**	64	4.9	Lack of broad-based agreement on reserve alternative by regional stakeholder group	No—state's preferred alternative did not meet the scientific criteria	8

* State (0–3 nautical miles) marine protected area designations include no-take marine reserves, state marine parks, and state conservation areas. State marine parks and state conservation areas (among other special designations) may allow some level of resource extraction and use.

** The South Coast region marine reserve designation is based on the "integrated preferred alternative" (IPA) the California Fish and Game Commission adopted in December 2010, in accordance with the California Environmental Quality Act. The development of the IPA was based on agreement by the region's stakeholder group. The IPA failed to meet the design criteria established during the Marine Reserves Working Group process and set forth by the Scientific Advisory Group in terms of the recommended levels of marine life protection. The South Coast total percentage and no-take marine reserve percentage include the state Marine Protected Area designations associated with the northern Channel Islands.

Certainly there were dramatic changes made by the state to the planning process in phase 2. The role of commercial and recreational fishery interests and stakeholders in the collaborative processes ensured that large networks of MPAs were not going to be created by the state. The levels of protection provided by the south coast mainland areas fall very short of the promise of the MLPA. A large percentage of the south coast region's MPA network is based on the prior designation of no-take marine reserves for the CINMS.

Few of the marine reserves off the coastal mainland extend to coastal processes, such as important coastal wetlands and estuaries. Coastal mainland reserves often allow some type of fishing activity. There is currently no deep-water marine area protected in MPAs off California. Birds, mammals, and pelagic species are not protected by the designated MPA networks in state waters off the coastal mainland. In addition, the level of biodiversity protection provided by the MPA networks of the south coast mainland cannot protect species or habitats from environmental fluctuation or climate-related disturbance events (Allison et al. 2003). The MLPA planning efforts show that there is a need to develop other tools, in addition to MPAs, to address the problem of the overuse of marine resources.

LESSONS LEARNED

A range of socio-ecological factors contribute to MPA planning and decision-making, including the general character of the institutional decision-making structure (e.g., consensus or agreement-based); the physical scale of the planning effort; the roles of stakeholders and scientists; the general character of the institutional culture(s); the values and interests of participants; the goals of the enabling legislation; and the political context, among others. The California MLPA process reflects a complex interplay between diverse stakeholder interests, values, and scientific information. The implementation phases of the MLPA show that as the scope of conflict between diverse interests expands, governments may attempt to reduce the conflict by limiting the range of issues in decision-making. Marine science shows us that large reserves are needed to protect a higher percentage of high-quality habitats to ensure that a greater number of marine species can survive. Yet, larger reserves will have economic impacts on fishers. Reducing the scale of reserve design may reduce these economic impacts. But the ecological consequences of smaller reserves need to be recognized.

In the long term, the socio-ecological costs of smaller reserves may be significant.

The MLPA decision-making process exemplifies a political process of negotiating ecology that includes the following general factors:

- Conflict is shaped by the physical scale needed to protect species and the level of resource use of marine resources.
- The larger the scale of protection needed, the greater the conflict between values and interests in a heterogeneous society.
- The more conflict, the more likely are government attempts to control conflict by "localizing" (e.g., reducing the scope of conflict to single issues, single sectors, or single species or places).
- The more conflict, the less likely decision-makers and stakeholders are to support the large-scale conservation measures that are needed to protect biodiversity.

My fear is, as we negotiate away the importance of large-scale protective measures, we also reduce our capacity to adapt to the major changes that are likely to occur to marine systems in a dynamic era of global climate change. Small reserves cannot protect marine systems from climate-related disturbance, or from other catastrophes, such as an oil spill. A major disturbance in marine ecosystems can degrade the habitat protected in small reserve. A large reserve can mitigate the impacts of climate-related disturbance and other catastrophes on habitat loss. Moreover, small reserves may protect some species, such as fishes and invertebrates, but they rarely protect migrating species or keystone species.

Given the ecological scale of our impacts to the sea, three potential outcomes of the process of negotiating ecology are:

- Government shifts the focus of decision-making from multiple species to single species (e.g., a shift from biodiversity considerations to fishery issues).
- Government shifts the focus of multi-sector or multi-scale governance to single-sector or single-scale governance (e.g., a shift away from integrated, ecosystem-based planning to a resource-based mentality).
- Government shifts the focus from multi-stakeholder decision-making to client-based decision-making that prioritizes economic values over conservation values.

Each shift reflects a change in the scope of participatory decision-making, and leads down a path of less conservation and preservation of marine habitat and species.

DOWN-SCALING ECONOMIC USE OF THE OCEAN

Without substantive change in value orientations and the approaches used to address the human impacts to coastal marine ecosystems, we will become more vulnerable and less secure in the face of climate change. While the designation of MPAs is needed, we need to consider a much broader set of planning tools and policy instruments to strengthen the conservation and preservation of marine systems. Ultimately, the protection of the world's ocean requires the cultivation of place-based ocean constituencies that can support a renewed maritime sensibility and ethos—not just embracing the economic or instrumental use of marine resources that are traded in a global economy but recognizing the intrinsic place-based values carried by healthy coastal marine systems. Maritime communities should be recognized as distinct places worth protecting. Protecting maritime place requires the preservation of marine ecosystems. In a context of global climate change and a global economy, a more ecologically integrative and holistic approach to resource allocation and biodiversity preservation is needed that can support particular maritime places and communities. Policies that support the up-scaling of marine life protection should be combined with down-scaling the economic use of marine species in global markets to ensure that maritime communities can adapt and be sustained.

Table 6.2 depicts a number of principles that can strengthen a place-based approach to marine life protection and a reduction of the scale of economic use of marine systems (McGinnis 2011).

Certainly the world's oceans carry life-giving values that are beyond an instrumental value orientation or "global development ethic"—for instance, a sea in a wild storm is valuable beyond the human capacity to understand it, while a sanctuary of shorebirds feeding in a coastal estuary embodies spiritual and sacred significance. The challenge is translating a maritime sensibility into governance and policy. This is not just an issue of science. Values matter. Our choices as consumers also matter. We need to gain a deeper understanding of the values of the seafood that we eat, where it comes from, how it is fished, and what the ecological consequences of commercial fishing are for ecosystems. In a characterization of the importance of wilderness values to marine

TABLE 6.2 BIOREGIONAL PRINCIPLES FOR COASTAL AND MARINE ECOSYSTEMS
AND PLACES

Type of ecosystem	Community with ecology *Restore the relationship between people and places*	Place-based economy *Restore the relationship between place and the mode of production and consumption*
Coastal	Promote place-based education and ecological literacy programs Protect rural lands	Use resources based on intergenerational values and equity and environmental justice Develop regional markets for regionally produced products Create value-added programs for products harvested sustainably
Marine	Establish large-scale no-take MPAS to protect marine ecosystems Protect keystone species, such as birds and mammals Promote ecological literacy programs	Develop regional fishery trusts and cooperatives promoting regional markets for local fisheries landed

preservation, Sloan (2002, 294) writes: "We should broaden social recognition of the intrinsic value of marine nature, exercise humility in the face of our ignorance, and clarify our faith in the potential payback from required abstinence. The first sacrifice is to have the appropriate sectors forgo economic benefits from resource extraction, such as a net reduction in fishing effort." Well-known author and marine scientist Carl Safina (2010) expresses a similar orientation:

> The main message from the albatrosses' realm is this: No place, no creature, remains apart from you and me—anywhere in the whole world. Seeing a parent albatross gagging up a toothbrush made me realize humanity has no borders. We've woven the albatross and many other creatures into our culture. That creates an obligation, and the opportunity to make a better world. We should do this not just for albatrosses, but also for ourselves. No less than a mother albatross delivering cigarette lighters and toothbrushes, human mothers pass toxic chemicals they've absorbed from food and water to their own babies in their milk. The albatross speaks to us of how much the world is changing, and also how similar we are. We are all caught in the same net of life and the same moment in history.

. . .

Hung out under bishop pine forests, with the wind blowing the
    seeds of the pioneer,
coyote brush;

to escape the fennel of the canyon, and to change a flat tire on the
   ridge-line road;
felt the cold down into my bones, thought of 20 warbler species,
   10 oak brothers, 4 manzanita; and
the notion of endemism, Noah's Ark, and oceanic moat,
   became clear . . .
In the midst of an old row of Father Eucalyptus
Cutting them down, followed a well-traveled path,
in the smell of animals, found my way home.

CHAPTER 7

The Challenge of Place-Based Ocean Governance in New Zealand

"If you took all of the genius that put a man on the moon," anthropologist and ecologist Wade Davis noted in a TED talk in 2007, "and applied it to an understanding of the ocean, what you would get is Polynesia."

Setting sail from Auckland, New Zealand, seven vessels (known as vakas or *wa'a*) and their sailors, from eleven South Pacific island countries, arrived in late June 2011 at the island of Oahu (figure 7.1). Each vaka was constructed from early drawings of a Polynesian sailing vessel. With colonization, the knowledge and practice of building and sailing in the old ways were forcefully separated from Pacific maritime cultures. But the seeds of this knowledge were not lost. Ocean, earth, and sky are bound together in these renewed voyages. Previous Polynesian explorers followed the path of birds, mammals, and turtles. In an effort to renew their maritime heritage, the sailors and navigators from eleven South Pacific Island countries set sail across the Pacific Ocean. With only the wind and stars to guide their journey, the crews on these vakas travelled thousands of nautical miles across the Pacific..

To celebrate the travelers' arrival, the Okeanos Foundation invited guests to participate in a three-day "kava bowl" ceremony in Oahu. Kava is a root harvested across Polynesia. When ground to powder it produces a mild sedative that is washed down with water or milk. The kava root is used by the Pacific Islanders for ceremonies to mediate conflict, and to foster conversation in periods of turmoil. It is one way

FIGURE 7.1 Vakas in Honolulu Bay, Hawaii. (Image: M. V. McGinnis.)

to share story, and to empower diverse voices in dialogue and decision-making. It is believed that the kava can "speak"—that over time, the kava ceremony can unite diverse maritime peoples under a common cause; and that the kava helps one listen to others and to nature.

The crews of the vakas joined in this kava ceremony. Each sailor was trained in the traditional ecological knowledge and sacred ecology of sailing by the stars and waves. Navigators learn a language of the waves and a knowledge of the stars that is shared during the Pacific voyage. Each wave speaks as it splashes against the bow of the vaka. Based on the sound of the splash, the sailor knows how close they are to the land, if a storm is coming, and the direction of the wind. Well over a hundred words were once derived from the sound of waves against the bow of the vaka. This knowledge of waves and the stars in the cosmos are the navigator's compass.

The importance of the voyage of the Pacific Islanders is the hope of regaining some of this knowledge and restoring the language of the sea; each vaka's voyage represents a return to the old ways of maritime knowledge of waves, currents, and winds. Each vaka is a response to the challenge that lies ahead for all islanders in a context that includes threats and pressures caused by the overuse of marine life and by climate change.

In the final day of the ceremony, a shared vision emerged that we are all on an island; we are all on one vaka. *Insula* is the Latin word for "island"; the insula is also a part of the cerebral cortex which is involved in consciousness and linked to emotion. To gain an understanding of our current socio-ecological condition we should imagine ourselves on a vulnerable island, a specific place of origin. In a deeper understanding of the sea, we can regain a consciousness and sensibility that connects multiple places. Maritime peoples share many problems—among them overfishing, pollution, coastal development, invasive species, and climate disturbance. If we begin to make connections across the Pacific, we can begin to create and imagine the modes of adaptation that are necessary to survive together as crew on a vaka in a turbulent storm. The storm of climate change is upon us. My fear is that we're losing our place-based and intuitive knowledge and communal maritime sensibilities, and as a consequence we are losing the capacity to adapt to our respective place. The language of the maritime (and rivertime) needs to be recovered and revered.

This chapter describes the challenge of ocean governance in Aotearoa (the four hundred islands of New Zealand). New Zealand is a vaka of origin and possibility, heritage and story, and a poetic, powerful metaphor of Planet Earth, reminding us that we are on an island of finite resources, floating in the sea of space. This chapter is a reflection of two years of research conducted in New Zealand, as a faculty member at Victoria University of Wellington, on the country's ocean governance framework. The ministries of New Zealand provided me with three years of support to study and make recommendations on how to overhaul their marine policy, and this chapter represents the major themes of this research. The chapter begins with a general overview of the changing socio-ecological context in New Zealand. It then provides a summary of the level of marine protection designated in the country. I also describe the case of marine aquaculture in New Zealand, and recent efforts to expand salmon aquaculture in sensitive marine areas. The future of the marine life of the Pacific Ocean depends on the country's commitment to protect marine ecosystems, which requires the integration of marine science, traditional ecological knowledge, and a renewed natural contract with the sea.

THE ISLANDS OF AOTEAROA

Māori arrived by vaka, with wisdom and knowledge of the stars, winds, and waves, and settled in Aotearoa. Centuries later, European explorers travelled by boat, and with new Western ideals also changed the natural world inhabited by the *iwi*. The maritime foundations of the diverse peoples of Oceania remain a driving force that is irrevocably embedded in diverse cultures and customs across the Pacific. And these customs are threatened. New Zealand has yet to protect the marine life that its diverse cultures are irrevocably connected to.

After its declaration of an exclusive economic zone (EEZ) in 1978, New Zealand, with a coastline longer than 19,000 km, has jurisdiction over 3 million km² of ocean. New Zealand's EEZ is the fourth-largest in the world, with an area of about fifteen times that of the land mass (or 5.7 percent of all the world's EEZs). This is an important point. A country of 4.3 million people is responsible for an incredibly valuable and extensive marine area that includes an array of coastal marine ecosystems. New Zealand's current ocean area jurisdiction spans more than twenty times the area of its land—1.2 percent of the surface area of the earth.

New Zealand is a Noah's Ark of species diversity. Its marine environment contains between a third and three-quarters of its endemic species. Many of these are unique to New Zealand. The high number of endemic species and the range of marine habitats associated with New Zealand make the county's EEZ one of the top hot spots for threatened biodiversity in the world. New species are found every year off the coast. Over 17,000 species of marine life have been identified in New Zealand's seas, including over 4,000 that have been collected but which have yet to be described. This comprises just over 30 percent of all known biodiversity associated with the country (Gordon et al. 2010). While the number of identified fishes has doubled over the past fifteen years, and is increasing at a rate of fifteen species per year, the number of undiscovered marine species in New Zealand waters likely exceeds the number that have been identified.

The marine ecosystems associated with New Zealand host an incredibly high level of native species diversity, including some of the rarest seabirds, whales, and dolphins on the planet. Almost three-quarters of the world's penguin, albatross, and petrel species, and half of the world's shearwater and shag species, are found in the country's islands and coastal areas. Many of these seabirds travel across the Pacific to feed. For instance, the sooty shearwater is the most abundant seabird off California, but it depends on the islands of New Zealand to nest. In addition, nearly half the world's species of whales and dolphins have been identified there, including nine species of baleen whales, seventeen members of the dolphin family, and twelve species of beaked whales (Gordon et al. 2010).

TROUBLED WATERS

In October 2011, an oil spill in the Bay of Plenty, along the east coast of New Zealand, intensified debate over the future of marine activities in its EEZ. An estimated 350 tonnes of oil leaked from the 775-foot vessel *Rena*, which struck the Astrolabe Reef in the Bay of Plenty. The vessel subsequently broke in two, and much of it is under water. Large numbers of containers were washed up on the shore, or sank. Well over 1,300 birds died as a result of the spill. But this number of marine life casualties is an estimate at best. (Note that recent scientific evidence indicates that between 600,000 and 800,000 birds were killed as a result of BP's oil catastrophe in the Gulf of Mexico. The original count was that roughly 1,400 birds were lost.)

The spill was New Zealand's worst environmental disaster in decades. It was a direct result and impact that occurs across the world ocean, as we develop the sea to transport goods and resources. As New Zealand continues to pursue marine resource development, a concerted effort to strengthen and improve the marine governance framework is needed. As chapter 5 noted, the "green" brand of "100% Pure New Zealand" is a double-edged sword—it represents an opportunity for the country to create the marine policies and programs that support the brand, and a vulnerability or liability with respect to the potential economic (and ecological) fallout if the country fails to live up to the brand. New Zealand's marine policy is far from "green," despite its reputation and brand.

Well over half of the country's EEZ has been leased for offshore oil or minerals development, and these leases did not require environmental assessment of constraints or mitigation for potential risks or impacts to marine life from minerals exploration or development. In addition, the country is a major exporter of commercial fishes landed in New Zealand from its EEZ and from other marine areas across the South Pacific. Unique and sensitive marine ecosystems, such as the Ross Sea of the Antarctic, are major fishing grounds for New Zealand's commercial fishing fleets. New Zealand's interest in fishing the Ross Sea is one primary factor contributing to the ecological decline of the Ross Sea, as it is overfished for its Antarctic toothfish, which is caught during the austral summer above the Antarctic Shelf and then exported and traded in international markets (e.g., to the United States and northern Europe).

Figure 7.2 depicts the synergistic and cumulative impacts and pressures of the multiple uses of coastal and marine resources. Human beings cannot manage ecosystems, but they influence ecosystem health with their activities and use of systems. The fundamental institutional challenge is managing human behavior. Management that is not integrative and holistic is bound to fail. For instance, if public policy is devoted to managing the particular use of a fished species without consideration of the much broader social, economic, and ecological context which the fish depends on, including water quality, habitat consideration, and other biophysical factors, the management of the fished species will likely fail. The challenge is to manage the use of the fished species with respect to the synergistic impacts of fishing and other human impacts on the fish. Ocean ecosystems are nonlinear systems; management of human behavior is difficult because of the functional, temporal, and spatial complexity of these biophysical systems.

Allison MacDiarmid and colleagues (2012a, 2012b, 2013) characterize the primary threats and pressures on the country's marine ecosystems and the loss of ecosystem services that are associated with these systems. These scientists used a model developed in the United States, now used by the United Nations Environment Programme. These important studies show that the two top threats and vulnerabilities stem from anthropogenic climate disturbance. By a considerable margin, the highest-scoring threat identified by the authors is ocean acidification, a consequence of higher CO^2 levels in the sea. The results indicate the scale of threats to New Zealand's marine ecosystems. The biophysical scale of global climate change, including the rise in sea surface temperature, increasing acidification of marine areas, loss of native species diversity, and other biochemical changes to marine systems, are threats that are well beyond the ability of New Zealand to address. Accordingly, to manage a fished species, it is important to consider larger-scale factors such as climate disturbance on marine habitat and species. One of the top-ranked vulnerabilities and threats is commercial trawling for fishes. Overfishing is considered a primary factor contributing to the disruption and degradation of marine ecosystems. Commercial fishers are likely impacting marine habitat areas, such as the benthic area, that are not well understood by scientists. The impacts of bottom trawling on New Zealand's benthic areas remain uncertain.

In addition to commercial fishing activities, the central government has leased major marine areas for prospecting for offshore oil and deep

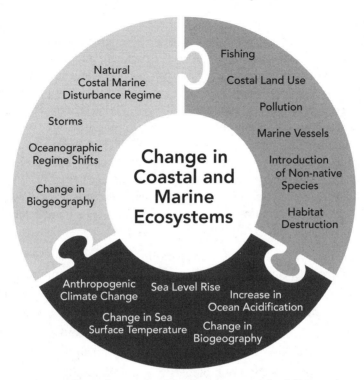

FIGURE 7.2 Change in coastal and marine ecosystems (after Miles 2009).

seabed mining. The largest area, and the one geologists say has the greatest potential, is the Great South Basin. The Crown supports the exploration of a range of minerals, including the extraction of deep-sea methane hydrates, the first such development in the world. Scientists have only begun to understand the ecology of the benthic areas of the country. Particular ecological concerns are associated with the impacts and effects of marine mining and offshore oil activities on the sea floor or benthic areas. Less than 5 percent of New Zealand's benthic area has been studied.

THE CHALLENGES THAT LIE AHEAD

One major challenge facing all maritime countries is a reliance on a sector-based approach to marine planning. A sector-based approach rarely captures the range of biophysical issues, interests, and values that are often associated with marine ecosystems. According to Lester et al. (2010,

577), "There is a historical legacy of piecemeal management that has largely focused on single sectors of activity and failed to consider marine ecosystems as interconnected wholes." Rosenberg and Sandifer (2009, 13) maintain that "under sector-by-sector management, trade-offs within a sector may be considered, but those among sectors are largely ignored and often remain unaccounted for." Likewise, Norse (2010, 184) argues: "This situation was hardly problematic when ample distance remained between swinging fists and noses, but in the face of today's increasing demands, a system of ocean governance less likely to give us healthy oceans and sustainable economies would be difficult to design. Without strong interagency coordination, sectoral management *cannot* work."

New Zealand's marine governance contributes to a number of challenges, such as:

- a spatial and temporal overlap of human activities and their objectives, causing user–user and user–ecosystem conflicts
- a lack of connection between the various authorities responsible for individual activities or the protection of marine ecosystems
- a lack of connection between offshore activities and resource use and the onshore communities that are dependent on them
- a lack of protection of biologically and ecologically sensitive marine areas.

Sustainable marine governance requires the institutional capacity to deal with socio-ecological systems that are complex, heterogeneous, dynamic, and prone to nonlinear and often abrupt changes (Young et al. 2007). There are synergistic and cumulative impacts from human use of coastal marine ecosystems, including the impacts of land-use activity, such as farming and ranching, on marine systems (Halpern et al. 2008). Marine governance in New Zealand is symptomatic of the problem facing most countries with coastal areas: ocean governance remains highly "balkanized" and often supports single-sector approaches to manage specific effects or uses. Governmental attempts to mitigate or adapt to particular resource uses with a sector-by-sector approach have proven to be ineffective and unresponsive to the cumulative and synergistic impacts and pressures from human activities.

Future marine policy in New Zealand will likely be based on how well the country resolves these tensions between the threats and pressures caused by the overuse of marine life, and the scale of marine ecosystem-based protection that is supported by the country. The relationship between the ecological scale of resource use and the scope of political conflict remains an important barrier to marine ecosystem protection. First, the existing marine governance framework is highly fragmented and is based on a sector-by-sector approach to marine resource use. There are eighteen main statutes, fourteen agencies, and six government strategies for marine planning in New Zealand (Vince and Haward 2009). Marine planning is currently based on a sector-by-sector approach to allocating marine resource use in New Zealand. Marine planning and decision-making are made more complicated by the fractured framework of laws, regulations, and practices that have been developed in New Zealand over the past thirty years.

Second, New Zealand is not meeting its international obligations when it comes to biodiversity protection. New Zealand has not created marine reserves within the EEZ that can protect ecosystems from human impacts. The protection of marine life is an important requirement in international conventions and treaties, such as the United Nations Convention on the Law of the Sea (Rothwell and Stephens 2010; Parliamentary Commissioner for the Environment 2011; Oceans Policy Secretariat 2003a, 2003c). Every coastal state is granted jurisdiction for the protection and preservation of the marine environment of its EEZ. For example, coastal states have the obligation to control, prevent, and reduce marine pollution from dumping, land-based sources, or seabed activities subject to national jurisdiction, or from or through the atmosphere. While New Zealand has access and right to use marine ecosystems of its EEZ, this use is predicated on the protection of marine life in accordance to international obligations. National policy that supports the value of marine biodiversity protection has not been fully developed for the EEZ, and the current marine reserve designations in the country fall short of international agreements.

Third, the country remains far behind the curve in international best practice in marine policy and ecosystem-based development and planning (Peart et al. 2011). Marine policies should be based on internationally recognized principles of marine life protection. An ecosystem-based approach to marine governance can contribute to a more comprehensive and integrated approach to marine ecosystem protection and integrated resource use across diverse management sectors.

THE POLITICS OVER MARINE PROTECTION

The scale or level of marine life protection in New Zealand remains an issue of debate that reflects different values and interests. The level of protection designated in the marine waters off the shore of the country should be recognized as an international concern, given the diversity of marine life associated with its islands. A range of planning tools and zoning strategies, including the designation of marine protected areas (MPAs) and other protection and management measures, have been used in New Zealand. Its Marine Protected Area Policy is currently one of the major drivers of biodiversity protection; it was developed in part to fulfill the country's obligations under the Convention on Biological Diversity.

New Zealand has created particular plans to protect species, such as listing of marine species as protected under the Wildlife Act. Other protection and management mechanisms can be or have been established in the EEZ, including marine mammal sanctuaries and fisheries closures, such as closure of seamounts and benthic protected areas. There are also other protected areas, such as the marine components of nature reserves. The country's Department of Conservation notes that 7 percent of the territorial sea and 0.3 percent of the EEZ is protected (McGinnis 2012b).

There is also growing uncertainty about the future of special-status species, including the Maui dolphin, the New Zealand sea lion, and other keystone species, which are some of the most threatened marine mammals on the planet. One issue of increasing conflict is the future of the Maui dolphin, one of the most endangered marine mammals of New Zealand. One of the country's leading marine scientists, Wayne Linklater (2012), notes:

> If we lose Maui dolphin it is likely that the effects will cascade through the food chain to radically change the community of plants and animals off our coasts. The loss of fish predators like dolphin can actually reduce ocean productivity for fisheries in the long-term. . . . We need to understand that the loss of dolphin can be a bad thing for the economy as well as a bad thing for the quality of our environment and our enjoyment of it. . . . The slowness with which the fishing industry and our political representatives act is a part of the problem.

Yet, the Crown in the summer of 2014 leased marine areas used by the Maui dolphin for offshore oil exploration.

Seabirds and marine mammals are especially vulnerable to many marine activities, such as offshore oil and gas activity and commercial fishing operations, and to climate-related disturbance. The scientific literature on the benefits of MPAs also shows that the expansion of reserve networks and other protective measures, including the use of marine zoning strategies, is needed as a climate adaptation strategy (Kingsford and Watson 2011). Scientists have shown that small marine reserves and other protective measures rarely protect keystone species, such as seabirds and marine mammals, which are vulnerable to large-scale changes in the marine environment.

The level of marine life protection provided by existing statutes and plans is also an issue of continued debate. A number of studies have noted that New Zealand has thus far designated less than 1 percent of its marine area as MPAs. Most of this protection exists in the Kermadec Marine Reserve and the Auckland Islands Marine Reserve; these two areas represent approximately 99 percent of the total existing protected area in New Zealand marine waters. The existing marine reserves associated with Kermadec Islands protect marine life and natural features within the territorial sea around the Islands. With respect to the benthic protected areas in the EEZ, scientists indicate that these areas are of low habitat value for biodiversity protection (Leathwick et al. 2008).

One challenge is the public's perception of the level or scale of marine conservation that exists in New Zealand. Eddy (2014) conducted a study of public perceptions of marine life protection in New Zealand and found a hundredfold difference between the perceived scale of the marine environment protected by marine reserves in 2011, at 31 percent, and the actual protection level of 0.3 percent. Part of the problem is what is meant by "protection." There are clear differences of perceptions between marine scientists and the members of the general public.

New Zealand needs to establish a representative network of no-take marine protected areas using new scientific information and technologies. Legislation designating a network of marine protected areas should supplement the further development of a more comprehensive EEZ governance framework. Marine areas that are essential habitat areas for nesting birds, for example, should be considered as refugia where marine activities, such as oil development, will not pose threats. Areas used by whales and other marine mammals should also be carefully identified and protected. Adequate buffer areas should also be designed and designated to further the protection of these important marine areas.

THE RACE FOR MARINE SPACE

Marine life conservation seems not to be a priority in New Zealand. A race for marine space and use is the current political and economic priority. Conflict between contending interests and multiple values associated with marine areas is shaped by two interdependent factors: the level of marine resource use, which often supports global trade, and the proximity and/or access of users to coastal marine areas. It is important to recognize that the scale and the scope of conflict often shape the politics of marine planning and decision-making (McGinnis 2012b, 2012c). For example, the scope of conflict is shaped by different political contexts associated with marine life protection; these include user–user conflicts (e.g., commercial versus recreational fishing interests) and user–marine ecosystem conflicts (e.g., fishery interests versus marine mammal protection). The larger the scale needed to sustain resource use and protect marine life, the more politically contentious decision-making and planning processes become (McGinnis 2012b).

The Case of Marine Aquaculture

A case in point is the recent call to expand marine aquaculture in New Zealand. Given increasing demands for seafood in international markets, interest in further expansion of the aquaculture industry has emerged in the country. In addition, new demands overseas for beef and dairy products have given rise to changes in land-use practices in the major catchments or watersheds of the country. Today, a country with a population of 4.4 million produces protein to feed 220 million (Hilborn 2011). The government's goals for growth in aquaculture include $1 billion in revenues by 2025, effectively tripling the current $390 million (Brownlee 2010). The current 5,700 hectares would be expanded to 16,000 hectares (McGinnis and Collins 2013).

The Aquaculture Legislation Amendment Act (No. 3) 239–1 of 2010 implements the national government's decisions on reforming legislation governing aquaculture. This act amends four separate earlier acts: the Resource Management Act of 1991, the Fisheries Act of 1996, the Māori Commercial Aquaculture Claims Settlement Act of 2004, and the Aquaculture Reform (Repeals and Transitional Provisions) Act of 2004.

Most aquaculture takes place within the territorial sea close to shore, and is an industry that can be influenced by terrestrial inputs, including water pollution. Terrestrial inputs may include sediments, nutrients, and other water pollutants that enter marine areas from rivers and streams (Gibbs et al. 2006). The industry relies on a marketing scheme that projects a "clean and green" image for the country's aquaculture products (Murray and McDonald 2010). For instance, farmed king salmon is sold to European markets, and is advertised as a product derived from the marine waters of "100% Pure New Zealand."

The expansion of any marine use should be based on a careful examination of physical constraints. Currently, there is a lack of scientific information on the coastal marine ecosystems of the country (Royal Society of New Zealand 2012). As the growth of marine industries is encouraged by the national government, there are few studies on the nonconsumptive values associated with the country's coastal marine ecosystem services (Royal Society of New Zealand 2011). The Royal Society of New Zealand (2012, 5–6) writes:

> For offshore resources there is limited information and monitoring about the nature, distribution, variability, and vulnerability of resources, species, habitats, and ecosystems. Therefore it is difficult to make sound resource management decisions in a consistent and evidence-supported manner without a great deal more understanding of the offshore environment. . . . Our limited capacity to carry out baseline research is only one example of the systemic lack of research, to the point where it is difficult to see how any marine resource can be effectively managed or any a balance struck between competing uses. Improving the capability to monitor cumulative impacts, assess the vulnerabilities of ecosystems, inform on ecosystem health and integrity, and to understand socioeconomic values of marine areas will help the success of integrated planning of marine resource management.

The allocation of additional aquaculture space within the territorial waters of New Zealand is complicated by the need to make decisions based on limited information. In a study on aquaculture in New Zealand, Banta and Gibbs (2009, 177) note that "regulators have often been forced to make resource consent decisions on relatively sparse information." Planning for the allocation of marine resources is also made more difficult by the diverse and often conflicting value orientations that are associated with the use of science and scientists in this policy domain, such as the conflicts among commercial, civic, and Māori epistemologies for marine stewardship. Indeed, there is a heated debate over the ecological and economic impacts of the future growth of the aquaculture industry in New Zealand. This debate includes issues of the role of science and scientists in aquaculture planning, and socioecological issues regarding the use of additional coastal and marine space for future aquaculture operations. Forrest et al. (2007) show that there is uncertainty in the level of cumulative effects or threshold effects for major aquaculture developments, such as finfish farms.

The Debate over Salmon Aquaculture in the Marlborough Sounds

New Zealand accounts for over half of the world production of king salmon. The New Zealand industry has grown into the largest producer of farmed king salmon in the world. The New Zealand King Salmon Company dominates the production of king salmon. There are no native or endemic salmon in the rivers and marine areas of New Zealand. The king salmon farmed in New Zealand originally derives from the McCloud River of the Sacramento Basin in California; it was introduced to New Zealand in the early 1900s. In addition, salmonids have been implicated in the decline of indigenous freshwater fish in New Zealand via competition and predation. These impacts are thought to have resulted in the fragmentation of some native fish populations (Townsend and Crowl 1991). Research shows that salmonids have altered the behavior and habitat use of indigenous species via competitive interactions (McIntosh and Townsend 1995).

Public perception of salmon aquaculture in New Zealand is generally negative (Shafer, Inglis, and Martin 2010). In an analysis by Banta and Gibbs (2009), social concerns during planning and decision-making processes were listed in 95 percent of the council's reasons for declining permits in the Marlborough District in 1995–2004. In general, conflict in aquaculture is likely to be higher where coastal areas are densely populated, such as near urbanized areas. Aquaculture is politically controversial because of conflicts between development and concerns over the ecological and economic impacts of the industry (Shafer, Inglis, and Martin 2010), the importance of preserving the natural character of ecosystem (Gibbs 2010), uncertainty over the ecological effects of marine resource use (Gibbs 2010), contention over the property rights regime (Gibbs 2010), and conflict between local residents, recreational users, and environmentalists (McGinnis and Collins 2013). The New Zealand King Salmon Company has proposed to make two changes in the Marlborough Sounds Resource Management Plan and apply for nine resource consents to farm in the area. While the majority of aquaculture planning decisions take place within regional or territorial authorities, the king salmon case was considered nationally significant by the minister of conservation and thus went to the Environmental Protection Authority (EPA) for decision. The king salmon case is considered nationally significant under the Resources Management Act (§ 143), based on factors such as:

- significant use of natural and physical resources (including significant volume of salmon feeds)
- effects on the Marlborough Sounds, a place of national significance with significant ecological, landscape, visual, natural character, recreational, and amenity values
- proximity to habitat of endemic and nationally endangered New Zealand king shag and Hector's dolphin
- likely irreversible changes to the environment (specifically regarding the fecal matter of salmon)
- expansion that is or is likely to be significant in terms of the Treaty of Waitangi—there are eight *iwi* that may claim customary interests where the farms are proposed
- potential to create positive economic benefits for the region
- widespread interest in the public regarding likely effects on the environment

The EPA Board of Inquiry leads a planning process that includes arguments for and against a proposal to expand king salmon production in the Marlborough Sounds, which covers 4,000 km² of sounds, islands, and peninsulas along the South Island's northeasternmost point. The company proposes that expansion of salmon aquaculture in this marine area is compatible with the maintenance of the coastal marine ecosystems of the area. Thus far, roughly two-thirds of the nearly 1,300 submissions to the EPA on the expansion plans oppose the proposal, and the planning process has been a costly one.

The Marlborough Sounds District Council notes that the proposed expansion of salmon farms conflicts with the conservation elements that are part of the resource management plan. The plan, completed in 2003, explicitly prohibits expansion of aquaculture in the marine area outside areas designated for aquaculture. The council maintains that the proposed expansion by the company overstates economic benefits and understates the ecological effects of the proposed new fish farms. A key issue of contention and debate is the potential ecological impact of salmon farming on the water column and the marine life of the coastal marine ecosystem of the Marlborough Sounds. The New Zealand Department of Conservation also has a role in the planning process, and believes that the proposed salmon aquaculture should carefully consider the cumulative environment impacts of existing and potential new farms. One impact is the increase in algal blooms that may result from the proposed expansion. According to the Department of Conservation, a range of other effects on the seabed, marine mammals, seabirds, recreation and tourism, economics, and navigation should also be carefully evaluated. The king salmon case highlights the tension between the drive for commercial gains in aquaculture and core ecological and social values for New Zealand's coastal waters.

One coastal management challenge in New Zealand is how to resolve spatial allocation for aquaculture and other coastal marine uses in such a way that the ecosystem services will be protected over time (Royal Society of New Zealand 2011, 2012). Conflict over spatial allocations will likely be based on diverse interests and value orientations. Likewise, as more treaty claims are settled, there will be an increasing need to include Māori values in resource management, such as freshwater and marine spatial allocation. The current coastal and marine management framework in New Zealand lacks an emphasis on integrative approaches to ecosystem-based planning and decision-making (McGinnis 2012a, 2012b). Commercial values, as dominant for aquaculture, tend to disincentivize information sharing and collaborative science. The current emphasis in New Zealand is primarily a sector-by-sector approach to spatial allocation for marine resource use and biodiversity conservation. A sector-by-sector approach has shown to be irresponsive to the management challenge of addressing multiple use conflicts over coastal marine areas (White, Halpern, and Kappel 2012). For instance, in a study of the benefits of a multi-sector and integrative approach to marine spatial planning (MSP), White Halpern, and Kappel (2012) contrast single-sector management with MSP by comparing the value of all sectors when wind sites are chosen based only on the value of the wind energy developments versus also incorporating losses to the other sectors. Moving beyond single-sector management to MSP can yield significant benefits through the strategic zoning of competing ocean uses by different sectors.

New Zealand would benefit from the adoption of a more integrative, multi-sector, and ecosystem-based approach to marine resource use allocation and biodiversity conservation. A unique approach will be needed for MSP in New Zealand, given the effects-based approach to marine management in the Resources Management Act. There is a burgeoning literature in support of MSP as a decision-making tool that can address the types of conflicts over values and science that are associated with coastal marine resource management (Ehler and Douvere 2009; Foley et al. 2010). The emerging conflict over the expansion of aquaculture in New Zealand's territorial waters is one example of the types of future conflicts the country will likely face as it expands marine resource use for oil and gas activity, renewable energy projects, and seabed mining (McGinnis 2012a, 2012b). MSP is characterized as a tool that can support integrative ecosystem-based planning (Douvere 2008; Gopnik 2008), and may be a tool that can assist coastal and marine managers in conflict resolution and joint fact finding in future resource allocation and biodiversity preservation. Science and scientists have a central role in MSP. A civic-oriented science serves as an example of how science can help address system-wide environmental problems and conflict between user groups. This approach to producing science reflects dialogue with community and collaboration across different disciplines.

Marine Ecosystem-Based Planning

The idea of marine ecosystem-based planning is generating considerable interest across the disciplines, and includes the use of new planning tools such as MSP, marine zoning strategies, and the designation of marine reserves. There is a burgeoning literature in support of MPAs and MSP as tools that can address intergovernmental fragmentation and conflict between contending interests and uses, and facilitate integrated strategic and holistic management across diverse sectors of marine areas (Halpern, Lester, and McLeod 2010).

National and international organizations and governments are realigning marine governance frameworks to reflect the values of the maintenance of ecosystem health and integrity, adaptation, sustainability, and precaution (McGinnis 2012b). These values are the new pillars of marine ecosystem-based planning. Coastal marine ecosystem-based planning includes a range of programmatic developments, including integrative marine policymaking, ocean zoning, large marine ecosystem programs, integrative coastal zone management, and MSP. National ocean frameworks in France, the United States, England, Canada, Vietnam, Japan, Australia, Brazil, China, Germany, Jamaica, the Russian Federation, the Netherlands, Norway, Portugal, India, Mexico, and the Philippines embrace these principles of marine ecosystem-based planning.

MSP can also be used in conjunction with MPAs and other planning tools (Halpern, Lester, and McLeod 2010). The promise of an integrative, ecosystem-based approach to MSP is that human beings can cooperate to plan for the large-scale spatial complexity and variability of ecosystems, and resource managers can resolve the inevitable conflicts between social, economic, and political interests that are often associated with marine spaces (Ehler and Douvere 2009). MSP can also support participatory and collaborative processes that can broaden the planning effort so that it is not limited to those who receive direct economic benefits from marine resource use.

One caution is worth noting with respect to the use of planning tools, such as MSP. Advocates of MSP often refer to appropriateness of land-use planning and zoning in terrestrial settings as a reason for the need and expression of the marine planning tool. There are problems in relying on terrestrial models of land-use planning; terrestrial models may be inappropriate to emulate in view of the dynamic scale and complexity of coastal marine systems. Oceans have very different characteristic scales (of function, time, and space) from terrestrial systems. For instance, the abundance and distribution of marine life is influenced by subtle changes in sea surface temperature and oceanographic processes, such as currents and eddies. Our human perceptions and values are based on the fact that we inhabit landscapes. Our understanding of the spatiotemporal features and processes of marine systems is poor, and often shifts over time with new insights into history, evolution, and scientific data (e.g., paleoecological, archeological, and ecological). It is difficult for us human beings to deepen our social, conceptual, perceptual, and psychological identification of what it means to live in the multidimensional and fluid medium of the dynamic and complex marine environment. Biophysical processes and conditions in the oceans fluctuate greatly at time scales from decades to millennia. The use of terrestrial models for marine governance warrants further investigation given the complexity and limited amount of scientific information on the natural history of marine ecosystems.

In addition, to be successful MSP should be more than a technical or scientific mapping exercise; marine ecosystem-based planning requires more than the formulation of zonal plans for particular uses of marine space. MSP is more than a bureaucratic or technocratic exercise. As a tool for decision-making and planning, MSP requires a strategic and forward-looking ecological approach to manage human behavior and the multiple uses of coastal marine ecosystems. As with all tools and technologies, the use and application of MSP may not represent an ecological panacea. There are pitfalls in the reliance on a view of MSP that deploys techniques to rationalize nature and to render the oceans predictable, to replace its self-sustaining, ecological function and structure with well-managed industrial, commercial, and recreational spaces or boundaries. While MSP may resolve the potential conflicts between the uses of coastal marine areas, ecological thinking is integral to the planning enterprise. Maintaining the life-giving values of coastal marine ecosystems will require that we overcome the limits of the "multiple use" mentality that is pervasive throughout government, which makes impossible a collective experience with the ocean.

Restoring Maritime Heritage

Marine governance depends not only on the capacity and ability of institutions to adopt new science-based planning tools like MPAs or MSP, but also on the cultivation of a broad ocean constituency in the public realm that supports a more sustainable place-based and ecological approach to planning, decision-making, and policymaking. This is where a hope for change resides. All inhabitants of Aotearoa arrived by boat or vaka. The Māori have inhabited Aotearoa for over 800 years. New Zealand's rich indigenous history, in combination with the maritime cultures of the country, represents a foundation for the establishment of a restored ocean constituency. Accordingly, translating the principles and multiple values that are associated with marine ecosystems into a comprehensive and holistic governance framework should be an important part of future marine planning and decision-making in New Zealand.

As the navigator on the vaka understands, the path of animals is one key reference home. Animals can be a source of inspiration and guidance. The voyage of the Pacific Islanders is akin to the path taken by *honu*, the green sea turtle. Based on the multiple sacred and spiritual meanings and totems, such as the salmon, swordfish, or in this case *honu*, we can find the hopeful connection between marine ecological spaces and cultural places. *Honu* signifies the death/renewed life processes held by indigenous Polynesian culture. *Honu* is a symbolic species that reflects the existing pressures that maritime communities, coastal urban centers, and diverse Pacific coast peoples and ecosystems are facing in an era of climate change. Thirty percent of the coastal beach areas that *honu* depends on for reproduction have been lost to increasing sea level rise and coastal development; and these same areas have been lost to unique cultures as well.

Honu is a spiritual and metaphoric guide that travels the world's oceans; a shared totemic emblem that symbolizes the diverse ways people and places coexist. *Honu* ultimately challenges our senses and our ability to think deeply, and to act with respect to a distant horizon. We can find that other animals provide a lifeline that reconnects people and places across seascapes. It is the journey and path of other animals that can ground an ecology of understanding, and a recognition of our shared fate. The oceanic path of *honu* reminds us of the cultural and ecological power of wildness, the fragile world we live in, our shared fate, and the human struggle to survive and adapt to the changes in ocean ecosystems.

To survive, we need to expand our sense of time and space to incorporate the needs of other places and other peoples in our daily lives. While we may be far removed from the islands of New Zealand, the fact is that how New Zealand protects the marine life of these islands will determine the fate of many species across the Pacific Ocean, and will influence the maritime cultures that depend on these species for survival. The level of marine life protection provided in New Zealand is not just an issue for the country's residents to be concerned about; it is a much broader concern, with global significance.

Historically, the geography of hope that led to the migration across the wild ocean to New Zealand is a shared value that is part of the country's rich and diverse maritime heritage. Policy innovation is part of the history of New Zealand environmental governance. Risk-taking, experimentation, and adaptation are required traits of island cultures. Today, the wild ocean is reflected in the brand of "100% Pure New Zealand"—a brand that kiwis embrace and is celebrated abroad. But as the spill of the *Rena* showed, it is a very vulnerable brand. Living up to the brand requires a renewed responsibility to live up to and adapt to this changing, life-giving blue planet.

CHAPTER 8

Toward a Blue Economy

Songs of Migration and the Leviathan of Global Trade by Sea

There is no folly of the beast of the earth which is not infinitely outdone by the madness of men.

—Herman Melville (1851)

Perhaps the war of the whales was inevitable. Perhaps the two most successful hunters on the planet were destined to collide. Humans had dominated the life on land for 150 centuries, while whales had held dominion over the world's oceans for 40 million years.

—Joshua Horwitz (2014, xvii)

We can sense the changes across the seasons by carefully listening to the places we inhabit. I can hear the change in the winds by listening to the sound of the cottonwoods along the Carmel River. In the early summer months, the chime of a southeast wind in the leaves tells me that a California eddy is blowing, and there is probably dense fog along the coast. In the winter, the sound of the south wind in the trees indicates that a rare Pacific storm from the Aleutian Islands is approaching central California. Compared to the winds of summer or winter, the winds of springtime create a different sound in the trees; a roaring rattle of leaves is followed by a chorus of birdsong and I see pollen in the air. Along the banks of the river, I see that the cliff swallows have returned, and they are collecting mud with their small beaks to rebuild their nests. Later in spring, the swallows will chase their offspring in the sky. When I see swallows I know the spring winds will blow, the fog may move farther offshore, and these winds contribute to upwelling conditions in the marine system. Upwelling brings the cooler water from the bottom up to the sea surface. With upwelling, there will be birds and mammals feeding on the krill and fishes that are attracted to these nutrient-rich waters. Winds and the different birds are subtle clues of seasonal shifts in the ecology of my place, but they are also signs of what is happening offshore. The winds blow, and I notice the seabirds flying in great circles, feeding on fishes. Gray whales are feeding as well, close to shore. Indeed, the migration of some whales indicates that the spring winds are blowing. In this sense, the sounds of the northwest wind in the trees along the creek's bank are subtle clues to the oceanic muse that is springtime for whales. When I hear the springtime winds along the bank of the river I imagine whales feeding.

Maritime history is in part a story of whales. In prehistoric California, the gathering of whales could be smelled far inland. Maritime native peoples created story, myth, and dance around the seasonal migration of whales; they celebrated the presence of whales. Imagine a not-too-distant past of the coast where beached whales were surrounded by California condors and brown bears. Flying from a mountain range to the coast in search for food, the condor was the first to find the carcass. In time, the great bears would migrate down rivers and creeks, perhaps following the scent of the carcass or following the condor, to the coast, to feed on the whale. Crows and other scavengers waited their turn to feast on the dead whale. The bones of whales were used by indigenous coastal inhabitants to create sacred maritime jewelry and other resources.

There are other stories to tell. One summer afternoon, I arrived at the beach to see a gray whale that had washed ashore. It was a beached whale that had been carried by the waves and currents to the shore. People had gathered earlier that morning and, to my surprise, some had carved out their initials or drawn faces and images on the whale's carcass. As if the flesh of the whale were a cinderblock wall in a ghetto, the whale's carcass became a canvas for urban graffiti. I saw "Bill loves Jill," with a heart around it, on the whale's tail. As the spectators stood in front of this whale graffiti taking photographs, I heard the faint cry of a child and the sound of waves crashing. I thought it was a monstrous act, and it is one reason I began to study the ecology of the Pacific.

Scientists arrived and with their saws and blades cut away at the whale's thick blubber to examine the inner organs and skeleton of the whale for some clue to why the animal had died. I sat on the sand and looked closely at the blubber that had been cut away. The scientists showed me that the bones on the right side of the whale were crushed—

they appeared splintered like broken tree limbs. Apparently, the whale was hit by a large vessel or a container ship, and had died from internal bleeding before washing ashore. A whale hit by a large vessel suffers great agony. It is not a quick death. The killing of a whale by a container ship reminded me of the irony of seeing a dead grizzly bear in Alberta, Canada, on the side of the road, that had been hit by a school bus.

After the initial autopsy, the scientists tested the beached whale's blood for pollutants and toxins. Some whales that wash up along the shore have significant toxins in their body. These toxins accumulate in whales over time; and whales can live well over a hundred years (while a container ship's life may be thirty years or so). During an autopsy of a giant bowhead whale from the waters off Alaska, a harpoon point was found embedded in the whale's neck. This harpoon point dated back to around 1880; this means that the whale was migrating in the Arctic waters during the Victoria Era.

Whales are an essential part of the marine food web. PCBs and other persistent organic pollutants appear in marine ecosystems; they come from a wide variety of sources on land and sea and can accumulate in the whale's blubber. The level of toxicity can reach a point where the whale is considered a Superfund site. If the amount of toxic material accumulated in a beached whale is insignificant, it can be transported to a landfill. In other cases, the whale is "disposed of" in the deep blue far offshore, or buried in the sand.

The dead whale is a message: there are no true "sanctuaries" that protect whales across their migration routes. A patchwork of marine parks, designated marine sanctuaries, and reserves have been designated, but a whale's migration cuts across these areas. Whales and other marine mammals are granted special protection status under the U.S. Marine Mammal Protection Act and the Endangered Species Act; but these mandates are difficult to uphold. We don't manage whales. The only way we can protect whales is through managing human behavior.

There are a range of impacts that we have on whales. We no longer hunt for whales off the coast of California, but few consumers understand that the transportation of their goods by large commercial ships impacts marine life. Our consumption drives trade by sea. Whale strikes are one impact among many that are linked to the new economic leviathan of the sea, container ships. The shipping industry remains one of the least regulated marine users, and in most cases and across many maritime regions there has been very little action by federal or state policymakers to address the threats these vessels pose to marine life.

Most policies are voluntary in nature, and incentive-based programs to encourage changes in shipping "behavior" have not been shown to prevent whale mortality.

One reason for the lack of federal action is that international trade is the foundation of the global economy. Roughly $4 billion per day of import and export value comes into the Port of Los Angeles/Long Beach. When you purchase a twenty-four pack of toilet paper, a tooth brush, a toy for your child, or clothing from Costco or Walmart, a whale could have been hit by a container ship carrying this product, or the noise of the ship could have changed the behavior of whale that were gathering near a port or harbor. We may no longer use harpoons to kill whales, but we continue to contribute to whale mortality as consumers of goods transported by a placeless global economy.

Gary Snyder (1993) poetically refers to the global economy as the "grinning ghost of the lost community." Maritime ecology requires a deeper appreciation of marine life, and a change in our consumptive patterns that support the global economy. This chapter's focus is on the impacts of large container ships on whales. The global trade by sea furthers the impacts of human activities on ecosystems and cultures, including the impacts associated with climate disturbance. The noise produced by marine vessels underwater can disrupt whale calls and change their behavior. Container ships and other vessels can hit whales. This chapter argues that one response to these contested modes of globalization is the creation of a blue economy and a maritime ethic of place. Protecting whales from large commercial vessels is not just a management challenge for policymakers or elected officials, nor is it an issue that can be resolved by the creation of new markets or the use of more appropriate technology by the shipping industry. Protecting whales requires an economic response that is place-based. The ultimate challenge of creating a place for whales requires that we rethink the global economy and change our economic modes of consumption and production to strengthen our connection to place while respecting the needs of whales and other marine life.

OF WHALES AND MEN

"Oh, the rare old Whale, mid storm and gale,
In his ocean home will be
A giant in might, where might is right,
And King of the boundless sea."

—Whale Song (from Melville 1851)

In *Moby Dick,* Herman Melville offers a depiction of whalers' greed and material pursuits while recognizing that whales are spiritual beings and transcendental attributes of oceanic wilderness—a place where a boat and its men become lost in the pursuit of fortune and revenge. Early maritime history included shipping goods, such as tobacco or slaves, and often describes the threat of hitting a whale at sea, often with catastrophic outcomes. But few whales could sink a vessel. Melville also builds on the biblical meanings of the whale—the sea leviathan and the human relationship to hunting whales embody our relationship with the sea as both good and evil. The creative act of Herman Melville's story casts a broad net of wild imagination into an ocean that defies human control and management—our drive to consume the whale's blubber for oil is suspended by the primordial mythologies of the whale, which have transcendental meaning. Ishmael, the story's protagonist, joins the crew of the *Pequod* to seek solace in the sea so that he can calm and heal the "November" in his soul. But he finds no cure for his sorrow at sea on board the *Pequod.* Instead, Ishmael finds great mystery in the spirit of the deep blue, and he describes Ahab, a heroic dark figure, with his passionate madness for revenge in pursuit of the great white whale. In some ways, Ahab's pursuit of the whale reveals the tragic sensibility of humanity's failed aspiration to control or overcome the forceful elements of a powerful living ocean.

Melville's message is a fundamental one: we cannot manage the whale. Whale behavior is relatively unknown. We can find transcendental meaning in Melville's depiction of Ahab's leg—the leg lost to Moby Dick is replaced by a carved-out bone of a whale. Ahab searches for vengeance against whales; but he has become part whale. The whaler at the top of the mast looks for birds across the blue horizon. The birds are attracted to the whale. The whale is a hidden island underneath the world, existing within the silent bosom beneath the seascape. Like Ahab, the *Pequod* (the whaling vessel) is composed of the whale, insofar as many of her missing parts have been replaced by whale bones. The distinction between whale and humanity is made more ambiguous by the whale's sinking of the *Pequod.* In the final passage of the book Melville invokes a harmonious unification of man, whale, and ship. Ishmael—as if he emerges from a biblical whale—is the lone survivor who is left to tell the story.

Melville wrote *Moby Dick* more than two hundred years ago. I have struggled to find meaning in his prose, and to translate this meaning to our *hunt by other means* for whales, not by harpoon but by our mere greed and consumption of goods delivered by container ships that cut across the sea, in the path of whales. The contested ecologies we presuppose that differentiate the whale song from our own song and heritage is a metaphor grounded in separation, alienation, and detachment from the sea. If we are to find a place to dwell and inhabit, we must respect the ecology of whales, their songs, and their migration.

The killing of whales for their oil and meat off the California coast ended by the early 1970s. With such loss and depletion of whale tribes by whalers, one hopes that the songs of whales are being relearned. But maritime song is being lost. The whales may be relearning their language. Our fish stories are now mere shadows of the *once* ocean. We cannot learn the language of whales; even if we try to listen and categorize their ways and behavior, we will fail to grasp their message. But we can respect them as different speakers who hold a profound knowledge of diverse maritime ecosystems, and respect them as different representations of the maritime. We hunted them, but they are the great hunters of the sea.

The large-scale sense of place that whales have is influenced by their respective underwater acoustic symphonies, which include songs, calls, and whistles that travel across the sea. Whales' migration, reproduction, and search for prey are to a degree predicated on learning and relearning new songs during their respective life cycles. The evolution of whales is a unique one: some aquatic animals moved ashore and lived on land, to become mammals; and then some mammals returned to the sea, to become whales. There is no specific migration grid or oceanic freeway for whales. Whale migration is adaptive; whales respond to the dynamic ecology of marine ecosystems, which is influenced by climate, biology, and oceanography. In their songs, whales call out and respond to other whale songs. Scientists believe that songs change with the dynamics of marine systems, the availability or distribution of prey species, and the climate-related and physical processes that contribute to biological productivity and that diverse whale species depend on to survive.

Early whalers followed the migration of whales. In following their migration, human beings were in many ways following the songs of whales. Several hundred years ago places like the Santa Monica Bay and Monterey Bay included great gatherings of whales feeding in the nutrient-rich California Current. These areas were major hunting grounds for whalers. European and Russian whale boats followed the path of whales up the Pacific coast from the lagoons of Baja, Mexico, to the northeastern Pacific and the Arctic Ocean. Various coastal places along Alta California recall this history of exploitation in a plaque or sign marking a landing area for processing the slaughtered whale's blubber. In the nineteenth century, whales were brought onshore near what is now the Marine Science Institute of the University of California, Santa Barbara. Off the shore of California, the whalers decimated populations of several species of whales, including thousands of gray whales migrating with their young from the wetlands of Baja, California, and humpback whales migrating from the tropical waters of Hawaii up to the cooler waters of the Aleutian Islands. Between 1905 and 1971, about 3,400 blue whales were killed off California (Monnahan, Branch, and Punt 2014). In the northeast Pacific, about 1,200 blue whales (20 percent of the population) were killed during a five-year period. In the 1960s, Soviet whalers illegally killed 500 blue whales off the West Coast. Whaling was not only the killing of whales but the silencing of the sea.

The impacts of hunting whales in other places were much more dramatic. In the ice ecology of the Arctic, the blue whale population suffered a 99-percent reduction in one generation of the life cycle of the species. If whale songs are learned, then whaling also had an impact on the songs (or language) of particular populations or tribes of whales. One can imagine that the knowledge of many whale songs was lost during whaling.

Some whale populations, such as the North Pacific blue whales, may be in a process of recovery to their pre-whaling numbers, but this does not reduce the need to protect the species from the multiple impacts of fishing, pollution, container ships, and climate change. In a recent analysis of data on the population of blue whales, Monnahan et al. (2014) note that the population has recovered to about 2,200 whales off the coast, which they argue is about the same population as before the whaling in the twentieth century. But this research is speculative at best. The science of marine ecosystems cannot tell us the natural baseline of a species. Marine science began well after the major predators of the sea were hunted and exploited. We do not have a reference point to some ideal natural state of the marine system—full of whales before we slaughtered them for their blubber—that existed in the past. Prehistoric hunting, for instance, had an impact on food-web dynamics and coastal marine ecosystems, and these hunting activities removed major predators from the marine system. We have few original or natural marine environments to study today. Accordingly, studies that propose a "rebound" in fish, mammal, or other populations of marine species should be critically examined.

THE SONGS OF MIGRATION

The whales turn and glisten, plunge
and sound and rise again,
Hanging over subtly darkening deeps
Flowing like breathing planets
in the sparkling whorls of
living light—

—Gary Snyder, from "Mother Earth: Her Whales" (1993)

Oral cultures have long learned the song of a landscape. Aboriginal peoples in Australia, for example, learn songs to literally bring forth the place where one is born and to acknowledge the presence of ancestors in the landscape. Place-based stories and songs were remembered for thousands of years, passed on from one generation to another. The landscape story embraced a deep meaning of kinship that also symbolically represented the more-than-human community. A place and the song are sung while walking the landscape; the song literally gives birth to the family. The song of a place, for indigenous peoples, is literally the place inhabited. And the songs of humanity that were place-based included the ways of other species. In prehistoric California, the languages spoken by 150 or more indigenous peoples reflected place and social identification with a common watershed.

Whale songs can also be understood as place-based reflections of the mammalian knowledge and understanding of the sea. While human beings depend on their sense of sight, whales and other marine creatures depend on hearing. The natural history of whales combines migration with song. Whales have cultivated their songs for millions of years. The diverse songs, calls, and whistles of whales are the foundation of their survival and sense of place. Whales make distinct calls that reflect their diverse populations and ecological contexts. Whale songs are a key component of scientific identification of the behavior and population of whales. For instance, Monnahan et al. (2014) identified two distinct populations of North Pacific blue whales—an eastern and western population—based on analysis of their respective calling patterns or songs.

Whale songs fill different niches in the marine system. The loudest animal on the planet is the blue whale; its call and song can reach 188 decibels and can travel hundreds of miles underwater. Relearning stories, whistles, and song are essential features of humpback whales, for instance, which learn new songs as they mature. Whales change the sounds and patterns of their songs throughout their lives and across generations. Humpbacks change their songs in response to other whales, and copy song elements they hear sung by other adult whales.

Blue whales often arrive in late fall off California. As baleen whales, the blue whales are attracted to the marine area to feed on the rich abundance of plankton that is often associated with deep, nutrient-rich waters located on the north side of Santa Rosa and Santa Cruz Islands. Primary producers of the marine system, such as plankton and juvenile fishes, are attracted to the upper water column, also known as the euphotic zone. This is where photosynthesis takes place, so a high level of productivity can result when the winds and physical processes of the marine environment are present. During their migration gray whales often follow the coast, and can be observed close to major points or bays. They are using these areas to avoid predators, such as orcas that are interested in their young. A delicacy for orcas is the tongue of whales. Humpback whales migrate from Hawaii to the Aleutian Islands and western Alaska. During their migration they come closer to shore to feed on fish or squid. These whales range from Mesoamerica to southern Alaska, migrating past California, Oregon, and Washington. Humpback whales are often observed in summer months off the southern and central coast of California.

Blue and humpback whales frequent the Santa Barbara Channel, and have been reported year-round within the vessel traffic scheme (VTS). But these reports are not reliable or credible indicators of the presence or absence of whales. Rather, they are a product of information gathered by whale-watching vessels that cruise the channel in search of whales. Information on whale abundance on the south side of the northern Channel Islands remains unclear. The National Marine Sanctuary Program is also working with other federal authorities to gain a better understanding of the abundance and distribution of whales through visual surveys from the air. In some areas, this information is available to ship owners to provide them with real-time data and maps of where whales have been observed. The VTS is a voluntary path for vessels transporting goods and containers north and south, usually entering the Port of Los Angeles/Long Beach. The U.S. Coast Guard plays a role in managing the VTS. But since it is voluntary, vessels can choose to take other paths. Vessels are not required to use the VTS of the Santa Barbara Channel. Also note that the VTS extends into the Channel Islands National Marine Sanctuary, which is home to a wide diversity of marine and coast-dependent species. One of the most important threats to island ecosystems is the impacts of vessels that run aground and hit islands, often losing their cargo and dumping their bunker fuel. Once a vessel has lost power, it cannot stop. Vessels may also stray off course and hit islands and coastal areas of the mainland. Vessel accidents have impacted island ecosystems across the world's ocean, and are a source of marine pollution as well.

WHALES AND THE ECONOMIC LEVIATHAN

Few large bays, estuaries, and lagoons along the California coast have not been dramatically transformed into major ports, marinas, and harbors. We have lost well over 90 percent of the coastal wetland ecosystems of California during the last hundred years. Ports are used for exporting and importing goods, and very little of what we consume is place-based. Rather, our goods come from container ships that travel across the world's ocean. Many of the laws and policies governing ports and maritime shipping are international in scope, and are developed by the International Maritime Organization. Various ports also have a port authorities, which manage shipping activities, including programs that address threats posed by ballast water and public health concerns associated with greenhouse gas emissions from ships. The laws and policies governing marine vessels across the high seas or within a country's exclusive economic zone are not the focus of this chapter. Rather, the focus is on the pressures of large commercial vessels on whales, and I provide a brief characterization of some of the policy developments to respond to these threats below.

The new leviathan is the large commercial vessel or container ship. Large container ships squeeze through the passages of canals or the narrow inlet between islands or channels to transport your hair spray, computers, phones, and food. Most goods imported into the United States travel by land or sea; less than 1 percent by weight of the total imports arrive by air. This is due in large part to the substantial cost of air freight, which can be sixteen times that of ocean freight. With respect to greenhouse gas emissions, maritime operators argue that trade by sea produces less emissions than other modes of transport. On average, goods shipped from China take around fourteen days to reach the West Coast of the United States or thirty days to reach the East Coast. Global trade by sea represents $2 trillion of consumer goods each year. We've grown increasingly dependent on large marine vessels for our global economy. Commercial shipping accounts for 90 percent of international trade. The West Coast ports of Los Angeles/Long Beach (LA/LB),

San Diego, the San Francisco Bay Area, Portland, and Seattle service a majority of all container ship traffic coming to and from the United States. Fifty-five percent of U.S. imports and exports depend on the Port of LA/LB complex.

There is a concerted effort by the maritime industry to build bigger ships that can carry more containers, to cut costs associated with fuel. Container ships are getting larger, and as a consequence major ports are expanding their capacity. By 2020, the capacity of the LA/LB complex will be expanded by 100 percent to serve the maritime industry. It remains unclear whether the increase in vessel size will result in fewer ships transporting goods. The size of container ships is based on the "twenty-foot equivalent unit" (TEU), which is the cargo capacity often used to describe container ships and associated container terminals in major ports and harbors. One TEU is the volume of the twenty-foot-long container or metal box that is the primary mode of transporting goods by sea. The average speed of container ships is about 24 knots (about 28 miles per hour). This may not seem fast, but if you are a sailor in a channel you had best sit tight and wait for container ships to pass before crossing yourself.

To save on fuel costs, ocean shipping lines can adopt "slow steaming" measures. Pioneered by Maersk Lines, slow steaming involves travelling at low speeds to reduce the amount of fuel consumed. In terms of operational costs, fuel consumption is a major expense for the maritime industry. Slowing down reduces costs for the shipping companies, but it increases the amount of time and resources it takes to move goods from one place to another. Many routes have seen transit time increase by between one and five days since the introduction of slow steaming. But, as I will describe below, slow steaming provides ecological benefits, including saving the lives of whales.

MARINE SPRAWL

The relationship between the container ships and whales is nothing like the battle between Ahab and Moby Dick. In this case, a whale will lose the battle. Whales of highly endangered species are killed by vessels across the world's ocean. In the Hauraki Gulf of the North Island of New Zealand, several Bryde's whales have been killed by vessels. Closer to home, whales are hit by large commercial ships within the California Current.

There is usually no physical harm to the vessel that hits a whale. Off the coast of California, ship strikes of gray whales are the most commonly reported, followed by fin, blue, humpback, and sperm whales. When large vessels such as container ships are involved, the ship's crew may be unaware that a strike has occurred. In *Ninety Percent of Everything: Inside Shipping, the Invisible Industry that Puts Clothes on Your Back, Gas in Your Car, and Food on Your Plate* (2013, 190), Rose George writes: "At sea, whales are bashed, battered, gashed, pinioned, and stuck. The true scale of trauma is unknown, like much else about whales. The size of the big ships means they can hit giant sea animals, lost boxes, or yachts and not notice. A bump, maybe, a faint jolt among all the pitching and rolling and jolting already. Visual evidence is rare enough to draw crowds."

Container ships pose a range of threats to marine life. There are no studies that examine the synergistic impacts of marine vessels on marine life. *Marine sprawl* is a relatively new term to describe the range of human impacts on marine ecosystems. The impacts of large commercial vessels contribute to marine sprawl in a number of ways:

- Vessels produce ambient noise and acoustic impacts on whales, often contributing to changes in whale behavior (Clark et al. 2006; Southall et al. 2007; McKenna et al. 2012a).

- At speeds greater than 11 knots, vessels are likely to strike and kill whales (Wiley et al. 2011; Silber et al. 2012; McKenna et al. 2012b; Monnahan et al. 2014).

- Vessels emit greenhouse gases that contribute to climate change (California Air Resources Board 2012).

- Vessels are a major contributor to water and air pollution (e.g., 45 percent of ocean pollution and 900,000 metric tons per year of oil pollution are linked to large commercial vessels).

- Vessels can have impacts on coastal and marine habitat due to vessel groundings.

- Vessels are the primary vector for invasive species in coastal marine waters.

There remains a significant level of uncertainty over impacts of container ships on whales and other marine life. Most research on the impacts of shipping activities has focused on whales and other marine mammals. Whales are vulnerable to collisions with vessels of all types, sizes, and classes throughout the world's ocean. Generally, the combination of high vessel traffic in whale feeding areas and migratory routes off the coast of California results in a marked increase risk of ship strikes and noise-related impacts for whales. The impacts include serious injury, mortality, or a substantive change in whale behavior. There is increasing evidence that the speed of large container ships does make a difference in terms of whale mortality (Silber et al. 2012), and that noise in the marine environment from human activities such as shipping can change whale behavior (Clark et al. 2009; McKenna et al. 2012a). The noise from container ships and other marine vessels arises when the force of the propeller cuts through seawater, which results in millions of voids and bubbles which then collapse to cause sounds underwater. These sounds compete with whale songs. Marine biologists have shown that human noises in the marine environment reduce mammalian vocalization, which contributes to changes in their foraging and breeding activities.

Estimates of the number of whale strikes and mortality levels are likely conservative. Many whales, including blue whales, sink to the bottom of the sea when they die, so we do not know the true mortality numbers from ship strikes. Between 1988 and 2012, there were a hundred documented large whale ship strikes along the California coast. At least eleven blue whales are struck each year along the West Coast (Monnahan, Branch, and Punt 2014). This is well above the number of whales allowed to be killed by marine vessels, which is 3.1 whales per year in accordance with the take permit granted by the National Marine Fisheries Service (NMFS) and in accordance with the Marine Mammal Protection Act. The NMFS has failed to respond to the killing of whales by container ships off the West Coast. In the fall of 2007 alone, four blue whales were struck and killed by ships and found in or near the Santa Barbara Channel and Channel Islands National Marine Sanctuary. Three years later, between July and October 2010, two blue whales (one a pregnant female that resulted in the loss of the fetus), one humpback, and two fin whales were found dead in and around the Monterey Bay, Gulf of the Farallones, and Cordell Bank National Marine Sanctuaries.

There are four designated National Marine Sanctuaries off California. In the most comprehensive analysis of blue whale movements, Ladd Irvine and colleagues (2014) tracked blue whales off the West Coast to better understand their habitat areas. These scientists tagged 171 blue whales off California at different times between 1993 and 2008, and tracked their migration and movement via satellite. They showed that many areas within nationally designated sanctuaries off California are hot spots for whales, which are attracted to strong upwelling areas that produce large amounts of krill, the main prey species for blue whales. In addition, Irvine et al. noted that the feeding areas of whales are also used by large commercial vessels out of ports from Los Angeles to San Francisco. The sanctuaries program has limited authority over a range of human activities within their jurisdiction. The lead agency remains the NMFS. Increasing threats from vessel strikes and other impacts of marine vessels remain an issue of debate and controversy. Sean Hastings, the resource protection specialist at the Channel Islands National Marine Sanctuary, says: "I can't have blue whales go extinct under my watch." Hastings has a story that is worth relating. In a Coast Guard helicopter, he flew over the mouth of the San Francisco Bay. From the helicopter, whales were clearly visible, feeding at the mouth of the bay. Also visible were several vessels, heading toward the Golden Gate and thus also toward the whales. He found himself screaming at the whales, "Get out of the way!" But often whales don't get out of the way.

RESPONDING TO THREATS TO WHALES

Information gathered by aerial monitoring, Automatic Identification System data on commercial vessel movements, krill abundance and distribution data, whale tagging, and passive acoustic monitoring can be used to prevent whale mortality. An analysis of ship and whale collisions shows that the chance of serious injury or death to the whale can be reduced by 50 percent at vessel speeds of 11.8 knots or less (Vanderlaan and Taggart 2007). Speeds less than 10 knots can reduce mortality levels by 70 percent. Time is money for vessel operators, and there are very few incentives to slow down. In addition, the logistics that goes into the planning of ship operations carefully considers the timing of the delivery of goods to ports. After containers are offloaded from ships, they are loaded onto trains or trucks for transportation across North America. Roughly two-thirds of the goods used across the United States enter the LA/LB port; most of these goods are then transported by greenhouse gas–producing trucks.

Most of the science analyzing the relationship between incidence of whale strikes and ship speeds has been done on the East Coast to strengthen protection measures for the highly endangered North Atlantic right whale (Wiley et al. 2011). The world's last 350 North Atlantic right whales live along the East Coast of the United States. These whales received protection from whalers in 1935. There remain two major threats to whales: vessel strikes and fishing gear. Ship strikes likely killed more than a third of the right whales reported dead between 1970 and 2007 (at least 24 of the 67 right whales reported dead during this period were killed by vessels). Right whales make calls over distances of twenty miles. These whales often leave Massachusetts Bay during the spring. In association with the bay, about 1,500 ship visits per year have been observed by the U.S. Coast Guard.

In hopes of detecting right whales of this region and the impacts of noise on their behavior, Christopher Clark, the director of Cornell's Bioacoustics Research Program, documents the presence of what he calls "acoustic smog"—the sounds of ships and other vessels passing his recorders. For whales, acoustic smog is a constant reminder of human activity; the noise drowns out the whale calls and songs. Whale behavior varies across species and marine contexts. Scientists have shown that across diverse marine areas acoustic smog can change whale behavior (Southall et al. 2012).

Studies of North Atlantic right whales show that noise pollution in the marine environment has reduced the ability of right whales to hear one another by 90 percent off New England. Mandatory measures were adopted by the NMFS in late 2008 to change the course of vessels when right whales are present after voluntary measures failed during the 1990s (Silber and Bettridge 2012). Mandatory strategies that support vessel rerouting have resulted in a decrease in ship-strike risk to whales (Silber et al. 2012). In addition, the NMFS requires vessels to reduce their speed to 11.5 mph in specific marine areas where the right whale is observed. (In other marine areas, it remains uncertain whether changing the course that vessels use can protect whales.)

Off the coast of California, we are only beginning to study the impacts of large commercial ships on whales. Megan McKenna of the Scripps Institute of Oceanography at the University of California, San Diego, has been studying the impacts of noise and whale strikes off Southern California for several years. McKenna and her colleagues showed that a voluntary speed reduction program that was initiated by the federal government to slow down vessels within the VTS of the Santa Barbara Channel did not work (McKenna et al. 2012b). Less than 1 percent of the vessels slowed down. In another interesting research project, McKenna et al. (2012a) describe one consequence of the economic recession during the last decade. Fewer container ships entering the Santa Barbara Channel was one result of the global economic recession, and this decline in vessel traffic benefitted whales. Hence, a slowing down of the global economy has an ecological benefit for whales migrating off Southern California. Given the failure of the voluntary program in the Santa Barbara Channel, in the summer of 2014 the Santa Barbara County Air

Pollution Control District, Channel Islands National Marine Sanctuary, Environmental Defense Center, and National Marine Sanctuary Foundation initiated an incentive-based program for the VTS of the Santa Barbara Channel to slow vessel speeds in hope of reducing air pollution and lowering whale mortality. To participate in the program, the ship needed to comply with four main criteria:

- transiting the Santa Barbara Channel between July and October 2014
- maintaining an average speed through the channel of 14 knots or greater for the past six months
- maintaining a consistent speed of 12 knots through this area without speeding up at other points in its route
- participating in the Port of Los Angeles and/or Port of Long Beach VSR program.

The program offered a financial incentive of $2,500 per transit. Based on the availability of resources, funding for sixteen transits was available. Verification of compliance was conducted by Automatic Identification System analysis and baseline data on fuel use from port to port to ensure that significant speed-up was not occurring elsewhere on the route. As of August 2014, six international shipping companies had agreed to slow down their ships in the VTS of the Santa Barbara Channel. The program received thirty ship transit requests, but did not have the support for this level of effort (see box).

The International Maritime Organization (IMO) is the primary organization that manages maritime activities. In December 2012, the IMO approved a vessel lane change for marine areas entering the ports of Los Angeles and San Francisco. The lane modifications were crafted by the National Oceanic and Atmospheric Administration, with support from the U.S. Coast Guard. It is important to recognize that a VTS is purely voluntary. Ships are not required to use a marine area's VTS. In addition, the IMO finalized new guidelines to reduce underwater noise from commercial ships. With respect to acoustic smog, in November 2013 the IMO adopted the following guidelines for large commercial ship operators:

- Recognize that shipping noise can have short-term and long-term impacts on marine life.
- Identify computational models for determining effective quieting measures.

Recent Emissions Reductions for Container Ships and Associated Consequences

Slowing down vessels is equivalent to removing 200,000 cars from California's roads, because slower vessels produce less pollution. Slowing down vessels also reduces public health concerns associated with these emissions. A number of new policies require a reduction of container ship emissions, including:

- The U.S. Environmental Protection Authority and the International Maritime Organization Emission Control Area took effect August 2012 and requires all ships to burn cleaner fuel.
- The California Air Resource Board has initiated a Vessel Speed Reduction Program to reduce pollution impacts of ships entering the Port of Los Angeles/Long Beach.
- The LA/LB Green Port Program provides incentives for vessels to slow down to 12 knots before entering the port to reduce air pollution. One consequence is that vessels take 1.5–2 hours longer to reach and leave the port. The port offers a suite of incentives (e.g. reduced wharf fees) and penalties. Compliance is reportedly over 90 percent.
- California's Ocean-Going Vessel Clean Fuel Regulation was adopted by the Air Resources Board in July 2008, and implemented in July 2009. The regulation requires the use of cleaner fuels by ship operators within 24 nautical miles of the California coast. This regulation led many vessel operators to choose not to transit through the established shipping lanes in the Santa Barbara Channel. About 50 percent of the vessel traffic moved out beyond the northern Channel Islands marine area, or to the south of the islands. These changes in the traffic pattern were driven by fuel costs and are increasing vessel traffic though the Point Mugu Sea Range operated by the U.S. Navy.

- Provide guidance for designing quieter ships and for reducing noise from existing ships, especially from propeller cavitation.
- Advise owners and operators on how to minimize noise through ship operations and maintenance, such as by polishing ship propellers to remove fouling and surface roughness.

These IMO guidelines, too, are purely voluntary (Leaper, Renilson, and Ryan 2014). The major challenge that lies ahead is to establish and require compliance, and then to monitor shipping activities to ensure compliance. Also, these guidelines do not address the noise from naval vessels and warships, or from other operations such as sonar. (The impacts of sonar on marine mammals are explored further by Horwitz 2014.)

BLUEING AN ECONOMY FOR MARINE LIFE

Are the toys you buy your children "whale safe"? Are the toothbrushes, umbrellas, and other household items that come to us from other countries contributing to the loss of great whales? Few of us recognize that our treatment of whales and other marine life is, in part, based on our economic relationships with one another and the choices we make as consumers of goods shipped to us by large commercial vessels. Our burgeoning global economic dependence on the transport of goods by sea threatens the future of the earth's whales and other marine life. In the late 1960s, the movement to protect whales within the California Current began with protests by Greenpeace activists. In 1969, a major oil spill in the Santa Barbara Channel galvanized a diverse grass-roots effort that led to the creation of the contemporary environmental movement. By 1972, a range of new state and federal environmental policies and programs had been created, including laws that require the protection of marine mammals. During the past decade, the gathering of whales has become a major economic draw in coastal towns that encourage whale watching. From the South Island coastal town of Kaikoura, New Zealand, to California's southern and central marine areas, tourists are in search of whales, not with harpoons in hand but with cameras and binoculars. The ocean is alive with whales "spy hopping," or racing one another along the offshore reef, to the cheers of spectators. I will never forget the sight, off Santa Rosa Island, of over a hundred California blue whales gathering to feed on krill.

The laws and programs that were adopted in the early 1970s did not prepare society to address the new threats to marine life caused by the increasing globalization of economy. During the last thirty years, container ships have been built to carry more trade, and there has been a dramatic rise in the number of these ships carrying goods across the ocean. There are new responsibilities of ecological stewardship that are profoundly based on the consumption patterns in our industrialized societies. The incorporation of ecological culture into everyday life requires a new mode of consumption and not merely new management programs or policies to address the threats and pressures that we have on whales. We need a renewed sensibility of place that expands to incorporate the ecology of others, our shared heart and mind.

A place-based consumer recognizes the spiritual and moral choices that drive appropriate economies of scale. Living with the sea requires a deeper understanding of how our consumption can impact other members of our maritime communities—whales, birds, turtles, plankton, and the blue horizon are part of our maritime sensibilities. We may refer to this as sacred maritime knowledge. There is a perceptual challenge to cultivating a greater respect and responsibility toward marine species. Human beings seem unwilling to reconnect their maritime economies and resource use with an oceanic heritage that cares for a shared more-than-human maritime community. We continue to count whales as if they were a "stock" to manage or use. A whale is not a resource or commodity. We rely on a relatively narrow field of vision and respect that is less open to the songs of whales.

The Practice of Blue Re-inhabitation

Ray Dasmann (1975) distinguishes between ecosystem cultures, whose "life and economics are centered in terms of natural regions and watershed," and biospheric cultures, which are directed toward global trade and exploitation of resources. Re-inhabitory people are committed to place-based economies of scale (Snyder 1996). A re-inhabitory maritime ethic is needed today. This ethic requires a renewed ecological sensibility that embraces a shared common identity and genealogy that cut across the landscape and seascape and link human activities with the needs of other species. They require that the choices we make as consumers includes the more-than-human aspects of our community, ecology, and economy.

A re-inhabitory maritime ethic acknowledges that the global scale of resource use cannot last forever; no food, water, or shelter is guaranteed. The Anthropocene age includes increasing vulnerability to changes in food and water security, diminishment of the availability of protein from the sea, decrease in the carbohydrates and cereals we need, and reduction of the availability and access to clean water. The practice of acting responsibly in this Anthropocene age requires that we choose to minimize our destructive impacts on ecosystems, oceans, and whales so that both ecological and cultural adaptation is possible for all life forms.

The original source of our cognitive capabilities is derived from the structures and organizational milieu of a shared oceanic genome. Our bodies are *of* the sea. Our "blue minds" are neurologically preconditioned to keep contact with the oceanic muse and maritime sensations. This is the case of our human genome whether or not you live along the coast.

Gregory Bateson was an ecologist, but he also spoke to the cultural sensibilities of place and nature in the cognitive processes of human beings. In *Steps to an Ecology of the Mind* (1972) he offers humanity a restorative song of the past—he encourages us to relearn our genealogical heritage and kinship with the more-than-human community. Bateson envisions a song that is relearned, as the great whales after being hunted are relearning their songs. It must start with an initiation of where you live, your sense of inhabitation, rekindled in a fireside chat or at the dinner table with family. When you swim in the sea, you are moved by a deep ecology of the sea. Your values are shaped by this interaction, because the sea has a language that can be learned and told. An ecology of the maritime, or rivertime, is a product of our place-based sense and our participatory mode of being in the world. A whale's song is a product of millions of years of participation and interaction with the currents, biology, and sensual characteristics of the sea. We share this heritage of song of place. New stories, art, theater, and maybe a carving from a whale bone can represent a new message to your family. As mammals, we can share in this divine reciprocity of the renewed sensibility and the perceptual modes of being in the world.

The careful reader has noticed that I often refrain from giving "the ocean" a smaller name—Atlantic, Pacific, or Artic, among other names for the sea. Naming the sea is a product of economic exploitation and the yearning for conquest. The naming of whales, mountains, islands, rivers, and other coastal places needs to be rethought. In the time of Pangaea, when the continents were intertwined like a Celtic knot, there was one sea, one ocean. As the continents divided, over the geological history of Planet Earth, diverse segments of this one ocean were separated, and species evolved accordingly. Geology was followed by biology. Culture should follow the lessons learned from ecology.

In this age of climate disturbance, for instance, the loss of ice in the polar regions is occurring much faster than expected. In the near future, a northern passage of container ships through the northern pole region is likely to develop. There is considerable interest in developing the resources of the Arctic. As the ice retreats, access to commercially valuable fishes, oil, minerals, and other resources seems likely. Recent scientific evidence shows that with ice retreat in the Arctic there has been an explosion of phytoplankton productivity in the marine area. As the northern passage develops, whales and other marine life will be impacted. With an increase in phytoplankton productivity, more baleen whales are likely to depend on this area as an important food source.

Toward a Blue Economy

Sacralizing the seascape is one response to the threats and impacts of marine sprawl. Early settlement of the first human inhabitants of North America followed the paths of *honu*, the green sea turtle, or the whales, to settle new places out of necessity. *Honu* represents a powerful sacred animal in the maritime cultures of Polynesia. Maritime peoples across the South Pacific are well aware of the cosmic sense of time and space, and the sound of waves lapping on their vaka's bow. For the indigenous peoples of the tundra and northern ecosystems of ice cultures, other animals carried symbolic importance for hunting, including the whales of the Arctic. Cognitive maps of hunting grounds remarkably resemble the topography of coastal places, and where animals could be found. Early hunters followed the paths and migration of polar bears and wolves across the frozen terrain. In California, bears followed the condor downriver to the carcass on the beach of a great whale. Early tribes followed the bear. Coastal trails of deer, bear, and other members of the community were well known.

To survive as an ecosystem person, taking responsibility for one's choices to consume is one key to the cultivation of a more resilient communal sensibility. Respecting the needs of whales is predicated on understanding the impacts of our consumer behavior and not so much on the bureaucratic management of vessels by federal or international authorities, such as the IMO. Finding refuge for whales is an issue that defies governmentality. There is no silver bullet. The major problem is our dependence on commercial shipping to transport most of the goods we use on a daily basis. This requires that we rethink and reorient our consumption patterns to incorporate a place for whales offshore to make sure that our products are "whale safe." We need to redesign our economic modes of consumption accordingly. We play a major role in global trade by the sea, and we are the fundamental drivers of the economic modes of transport. The place of whales needs to be better integrated into the economic scales in which we produce and consume. On the shores of the continents of the world, ports and harbors have been built to expand the globalization of goods, and this has not been without major ecological consequences for the climate, currents, and marine life that both whales and humanity depend on.

During the past decade, there has been an interest in developing a "blue economy." For some, a blue economy means development of the commodities that may exist in a marine system. A blue commodity is a source in the ocean redirected for human use. It could be the wind or current redirected for energy to serve the needs of a coastal town. It could be mining with new robotic technologies or trawling the sea bottom based on a quota system that allocates commercial fishes. This instrumental meaning of a blue economy perpetuates a utilitarian principle of conservation; it encourages development and growth that is measured, strategic, considers environmental constraints and impacts, and is based on faith in technological and managerial control that fosters resource use across time. To a large extent, the focus on technological fixes and managerial control to address the impacts of commercial shipping on whales is an example of this type of blue economic development. More efficient shipping can reduce the costs to the industry while reducing the threats and impacts to whales. Slower ships reduce public health concerns related to greenhouse gas emissions and can reduce the mortality levels for whales. But these responses to the pressures and impacts of large commercial ships will likely not be enough. This meaning of blue economy is primarily anthropocentric insofar as it serves the economic needs of human beings, and is grounded in a faith in technological efficiency and market-based responses to the pressures of marine sprawl.

A deeper meaning for the blue economy is one where *blue* is used as an adjective (like *green*). But *blueing* an economy is much more than greening capitalism, technology, and society. A truly blue economy is much more than promoting natural capitalism and creating renewal energy offshore. Blueing an economy requires a depth of perception—a deeper appreciation, respect, and responsibility for maritime cultures and peoples, and the ecosystems that we are irrevocably connected to. The answer to our maritime problems cannot be found in more efficient markets or in some form of appropriate technology. A blue economy seeks to preserve and restore the local, place-based maritime economy—this necessarily extends the values of consumption and production to include the more-than-human maritime elements of place. Marine life is more than a bundle of resources to be sustainably managed or exploited.

We need a blue worldview that embraces the connections, linkages, and relationships that exist across maritime space to embrace the ecology of whales and their songs. A blue economy is place-based—self-organizing, self-generating, and self-renewing—and supports a mode of consumption that recognizes the limits of ecological scarcity and the needs of other species.

CHAPTER 9

Islands in a Turbulent Sea

He wa'a he moku, he moku he wa'a.
The canoe is an island, the island is a canoe.

BUILDING THE CANOE

It is impossible to carve out a canoe without a deeply held acknowledgement of the ocean power of the shared breath (or *hā* in Polynesian for the "breath of life") and vernacular understanding of the sea (Grim 2001). On the coast, I once watched the last canoe builder of a tribe in the Solomon Islands at work. He found the wood he needed floating in a lagoon and dragged it to shore. He worked for hours with his hatchet cutting into the log, then shaved the sides while always measuring for the depth and width of the canoe. The final act was to put the canoe's hull to fire. After the wood was carved, the fire cured and strengthen the hull of the canoe. I watched the fisher use the canoe many weeks later.

The canoe's sail captures *hā*, and the sail and the canoe are shaped by the hands and tools of canoe builder. These tools and the knowledge of canoe building are also part of the tapestry of *mana* (power). To build a canoe, an interdependent sense of *mana* is also believed to be an essential part of maritime hunting, sailing (and how to set sail), fishing (and what fish hooks to use), and place-based living (Firth 1940). For Polynesians, the ocean is the embodiment of *mana* and can only be respected by understanding coastal and marine processes such as the currents, winds, and waves. *Mana* has been translated as supernatural power, a psychic force, magic, a totemic principle, a divine force, and a universal and wonderful force of nature (Firth 1940, 484). *Mana* is a shared power that links a culture to the sea and place. It is not based on a constructed sense of privatized use of a marine resource or the social contract of private property that grounds the Western philosophies of nature as a commodity or a mere object of instrumental value. The transpersonal beliefs that grow out of *mana* extend from the land to the sea, from the community to nature and the more-than-human world.

The navigator of a vaka uses the stars and a profound sense and understanding of the sea to sail across the Pacific Ocean. The communal knowledge systems of canoe building, fishing, and maritime navigation took thousands of years to develop. Knowledge of where to fish, how to fish, and when to fish was as important as the intergenerational use of carving tools employed by the vaka builder. This knowledge is a common trait of most indigenous maritime peoples that was passed on, and celebrated in story, song, ritual, dance, and ceremony. The knowledge of the navigator, fisher, and canoe builder is represented in the communal relationship that is taught in stories of *mana*. The marine area is perceived as an extension of the land and coastal area—there is no separation of the land from the coast and the sea.

For the Māori *iwi* (tribe) of the northwest coast of the Northern Island, the return of the sacred fish the blue moki was indicated by the presence of a particular star in the night sky. The first moki that were caught were offered to the maritime elders as a gift of reciprocity and out of respect to *mana*. Gift giving is essential to the cultural tradition of maintaining *mana* in the maritime community. The elders of the *iwi* hold the *mana*. If the principle of reciprocity and gift giving was violated, no fishing would take place and no fish would be found. The presence of moki was a gift from the sea. Both fisher and moki are interdependent aspects of *mana*.

In the Solomon Islands, carvings of marine life are reflections of an intimate and expanded sense of kinship and community linked to the sea. A great circle of marine species carved out of wood begins with the children in the family collecting shells that will be used in the inlay of the carving and sculpture. The wife prepares the wood and plays a crucial role in preparing the inlay. Ebony, rosewood, kou or kerosene wood, pale sapwood or dark hardwood is carved with mother of pearl and nautilus shell used as inlay. The husband carves the circle of marine life using tools and knowledge passed on from his father. The carving of marine life and other sacred materials represent an extension of the family's knowledge and the household respect for the sea. But it is more than respect, the carving is the embodiment of the *mana* shared by the family and the place. Carvings and canoe building reflect beliefs about the origin of all life forms and are linked to certain archetypal themes. The carving is a totem of connectivity, power, and reverence for a living ocean. The sea is the breath of the family and community. It is shared in the greetings between tribes and family, and in the winds that are captured by the sail.

ISLANDS IN A TURBULENT SEA

The loss of island cultures and ecosystems is a dramatic symbol of the loss of place and the ecological sensibilities that have long shaped maritime society and nature. The island peoples depend on the Pacific Ocean (see box) for survival. The coastal landscapes and marine areas that people depend on are increasingly threatened by the hurricanes, typhoons, and rising seas of climate disturbance. In a biblical sense, Noah's Ark symbolizes a collection of Pangaea's biodiversity placed on a boat to survive the mythical flooding of the planet. There is no Ark or final refuge that can protect the existing cultures of the ethnosphere or the great circle of animals, plants, and insects that have evolved during the last evolutionary epoch. For several coastal regions and island peoples, the Ark is indeed sinking. There few lifeboats to depend on, and there may be no safe refuge.

Climate change will render entire coastal and island communities uninhabitable, and there are few international frameworks, conventions, or agreements that support the rights of "climate refugees." Rather, a focus is on large-scale industrialization and continued reliance on destructive fossil fuels. Arctic peoples and the native inhabitants of small islands have a fate in common—many of them will lose their place, community, food security, and home from the multiple pressures and threats posed by climate-related impacts, including king tides, coastal erosion from major storm events, and the impacts of hurricanes or typhoons. The melting ice of the northernmost island of the Arctic raises the sea level, which in turn threatens many of the low-lying Pacific islands, atolls, and coastal areas. Marine life is also changing in accordance with the climate. The Arctic ice once prohibited the movement of species between the Atlantic and Pacific Oceans. But this is a relatively recent phenomenon in the evolution of the planet. Species were isolated and evolved accordingly because of the ice barrier associated with the Arctic. With climate change and the 50-percent decrease in ice in the Arctic that has occurred we are returning to a single ocean where the

Characteristics of the Pacific Ocean

- The Pacific Ocean covers one-third of the earth's surface area.
- Of the Pacific Islands region, 98 percent is ocean, 2 percent is land.
- Pacific Island countries and territories have a total exclusive economic zone of 38 million cubic kilometers.
- Ten million people live in its twenty-two island countries and territories.
- Pacific Islanders are the most culturally diverse peoples of the world, with 1,500 languages spoken (25 percent of the world's total spoken languages).

Pacific and Atlantic Oceans are connected. Once isolated marine mammals are interbreeding with Arctic species; the Arctic ice retreat is increasing hybridization of marine species. As the Arctic glaciers melt, we can expect 1.5 to 3 feet of sea level rise by the end of this century, which will likely inundate a hundred million homes and hundreds of islands and coastal areas. Climate change and sea level rise will force thousands of island peoples to evacuate their maritime places. This forced migration of people will be more dramatic if the ice of the Arctic and Greenland melts entirely. In such a scenario the sea may rise more than 22 feet. Climate change is most dramatically represented by the major changes in Arctic sea ice that have occurred during the past several decades. Low-lying small islands, atolls, and coastal areas will suffer the consequences of sea level rise, ocean acidification, and the loss of coastal marine habitats and species.

Pacific Islanders are among the most likely to be affected by climate change, a rising sea, and storms. Twenty-two island states across the South Pacific contain the world's highest proportion of native species diversity per unit of land area and over 600 indigenous languages (Egan 2008). These islands have a combined population of 10 million people, who contribute a minimal level of global greenhouse emissions while sharing a disproportional impact from the more populated, industrialized economies. Between 665,000 and 1,750,000 climate refugees could be forced to migrate and relocate across the Pacific region by the middle of this century (Campbell 2010). At stake is the potential loss of myriad indigenous maritime knowledge systems that have been passed down over generations (Adger et al. 2011).

This is particularly the case for island cultures and coastal societies, which are some of the most threatened languages and ecosystems on the planet (Whittaker and Fernandez-Palacios 2007; Hong 2011, 2013; Hong et al. 2013). We may lose roughly 80 percent of the indigenous spoken languages by 2020. Many of the languages that will likely be lost are those spoken by coastal and island peoples of the Pacific (Hong et al. 2013). The socio-ecological diversity of island peoples is threatened by increasing cultural homogenization, economic globalization, and impacts associated with global climate disturbance (Maffi and Woodley 2010). The globalizing forces of economic exploitation (e.g., the exchange of human and natural "resources") and climate change contribute to the decline in ecosystems services and increasing ecological insecurity. Ecological insecurity is rapidly becoming a transboundary challenge that links industrialized and less developed economies and cultures. The poor and less developed are more vulnerable and at risk than those of industrialized economies, and they do not have the necessary resources for the types of socio-ecological changes that are emerging in this Anthropocene era.

FORCED MIGRATION AND RELOCATION

Relocation—whereby livelihoods, housing and public infrastructure
are reconstructed in another location—may be the best adaptation
response for communities whose current location becomes uninhabitable or is vulnerable to future climate-induced threats.

—Robin Bronen (2012, 17)

With respect to the impacts of climate change on island peoples and places it is not so much a process of recovering a sense of place or a community-based ecological sensibility as it is first grieving the loss of place, and finding a new home to migrate to. Relocated maritime communities face difficulties in their new settings. This is particularly the case when relocated peoples are immersed among members of different cultures. Tension and conflict often develop between the divergent communities of those who are relocated and the old communities. There has been over a hundred years of relocation of Pacific islanders. In Michael Lieber's edited compendium *Exiles and Migrants in Oceania* (1978), a number of case studies are devoted to the challenges of relocating island peoples during early colonization. Relocation has been part of the colonizing force of Europe, Japan, Australia, and the United States. There have been a range of reasons for relocation in the past—among them volcanic eruptions, drought, population pressure, nuclear testing, conversion to Christianity, tribal conflict, tropical cyclones, coastal development, tsunami, war, mining, and nuclear contamination.

The process of finding safe refuge in a new area is made more difficult by the lack of collaboration and coordination on the part of industrialized countries to responsibly agree to protect these island peoples (Hong et al. 2013). Rarely does a relocation program enable the preservation of the unique cultural heritage, language, and knowledge-based systems of displaced peoples. Climate change renders entire localities uninhabitable, and the loss of maritime bioculture is inevitable (Bartlett 2002; Brown 2007/2008). The most extreme case of climate injustice is not only the likely loss of the ecosystems that local people are dependent on but the loss of place and identity that is likely to occur. The U.N. Human Rights Council has issued several resolutions that acknowledge that climate change is a threat to human rights. Underlying these resolutions are principles of climate justice that have only begun to be explored by academics and policymakers. While the scientific impacts of climate change on coastal and marine ecosystems have been an issue of serious deliberation and evaluation, there has not been a comprehensive approach to respond to the many challenges of relocation and displacement. The biophysical sciences have offered a range of vulnerability and risk assessments at different scales and across a range of locales. A number of plans for island relocation are in progress. Yet, the cultural dimension of lost maritime identities remains underdeveloped at best (Adger et al. 2011).

Inhabitants of low-lying coastal areas and small islands, such as atolls, will face *forced* migration. In 2005, the inhabitants of Vanuatu Island become the first people to be relocated because of climate change. Over a hundred villagers from the Lateu settlement were moved to the island of Tegua, in Vanuatu's northern island chain. Soon after, inhabitants of the Carteret Islands (Killnailau in the native language) of eastern Papua New Guinea were relocated. The people of Carteret have lost more than 50 percent of their land since 1994. In 2005, the Papua New Guinean government called for the evacuation of the small atoll of Carteret, making this island among the first of what is likely to be a long list of hundreds of islands whose inhabitants will be climate-forced migrants. In 2009, the first group of Carteret Islanders were relocated to Tinputz on Bougainville Island, a site allocated by the Catholic Church.

Fiji is active in the relocation process, with over thirty coastal communities recognized as threatened (Secretariat of the Pacific Regional Environment Programme 2014). A range of adaptation plans have been developed by the Fijian government. Most of these plans include some financial resources and technical assistance to relocate a number of remote coastal villagers to other areas (Meakins 2012). For example, the village of Vunidogoloa was the first entire Fijian village to be relocated, in 2014. The village was relocated from Natawa Bay, the largest bay in the South Pacific, to their new home, which they named Kenani, which is Fijian for Canaan, the promised land. The cost of relocation was about one million Fijian dollars (almost 500,000 U.S. dollars; Secretariat of the Pacific Regional Environment Programme 2014). This included construction of new homes, farms, fish ponds, and copra dryers. In the future, coastal peoples from nearby islands and other neighboring island nations, such as Kiribati and Tuvalu, may be relocated to Viti Levu (Davenport 2014).

The preservation of local and traditional knowledge, language, and epistemology should be an essential part of a climate relocation program (Knodel 2012). We need to respect the *mana* of the sea and coast, and build a new canoe that can capture the human imagination. We need to gain a deeper knowledge that can reconnect our lives with the sea, which contributes most of the oxygen we depend on to maintain life. The canoe's sail captures the wind no less than life on earth depends on the oxygen produced by plankton. Relocation of traditional maritime peoples requires much more than developing mitigation and adaptation plans. One challenge of the relocation of people who are suffering the loss of their place is to maintain the language, cultural knowledge systems, and self-sufficient economic lifestyles that have existed for thousands of years.

MOVING TOWARD A THEORY OF BIOCULTURAL JUSTICE

Garrett Hardin's (1974) "lifeboat ethics" is one response to this uncertain future of global ecological insecurity. Hardin argues that those in favor of a principle of justice support, to borrow the phrase from Herbert Spencer, an "ideology of the minnow." Hardin describes a lifeboat in a turbulent storm brought on by overpopulation and resource scarcity. The lifeboat has a limited capacity; there is only so much water and food available to those who are lucky enough to swim and climb on board the lifeboat. For those on board, Hardin maintains that it would be unwise to let others on the lifeboat because there is not enough food or water for everyone. To address the problem of resource scarcity and overpopulation, Hardin supports a form of centralized authority to reject the temptation to let others on board. A principle of justice that supports a broader allocation of resources (e.g., to less developed countries) will eventually lead to the sinking of the lifeboat.

Hardin's analysis seems drastic and draconian, but the global *realpolitik* is that the life boat motif is the prevailing intergovernmental response to the biocultural impacts and vulnerabilities brought on by climate change. The climate refugee is placeless, a passenger on a sinking ship without a lifeboat. Existing international human rights agreements and conventions do not protect the communities that will need to relocate because of climate-induced changes to their home place. The 1951 United Nations Convention relating to the Status of Refugees does not give those displaced by climate change the legal status of "refugee," so no state is required to accept them (Knodel 2012). A human rights program at the international level is needed to assist people who are forced to migrate because of climate change. Such a program must ensure that human rights (e.g., the right to clean water, the right to maintain biocultural heritage, the right to food) is extended to those living in maritime and coastal communities, and that adequate sanctuary is provided to climate refugees.

The life boat ethics described by Hardin fails to recognize that there are benefits to be found in an expanded theory of justice that can protect the diversity of other maritime cultures and ecosystems. There is need for a broader theory of justice that can embrace other forms of knowledge and life in an age of climate change. A third metaphorical boat is needed today: *we are all on a canoe sailing in a turbulent sea, and the storm of climate change is upon us.* My fear is that we're losing the place-based intuitive knowledge and communal sensibilities that are associated with indigenous maritime communities. With the loss of these cultures and communities, we also lose the knowledge systems that have long played a key role in supporting adaptation to previous climate-related changes.

Over the past several decades scholars have focused on the development of a theory and practice of justice. As we recognize the importance of biological diversity linked to cultural diversity it is important that we begin to expand the theory and practice of justice. With respect to the impacts of climate change, David Schlosberg (2012, 457) argues for an expanded epistemology of justice that can protect and maintain place-based peoples and their "natural and social worlds." He writes: "The vast majority of the current theories of climate justice are focused on frameworks of prevention and mitigation, or on the distribution of costs of adaptation to climate change" (446). In the context of forced migration and relocation of island peoples, a deeper socio-ecological approach to climate justice is warranted. Schlosberg argues for an approach to justice that recognizes the "basic needs" of local people, not merely the rights of a community who are at risk from climate vulnerabilities at the community level. In other words, the basic-needs approach proposed by Schlosberg requires a theory and practice of justice that are more than establishing an international framework agreement or convention that supports human rights for climate refugees or the allocation of resources (e.g., financing relocation of a community) to those who receive the burden of the risk or cost from climate impacts. The practice of justice in this age of climate change should necessarily combine and integrate a biocultural dimension across location and scale. Schlosberg's emphasis is on maintaining not just the ecological foundations of local places and communities but also the cultural and epistemological dimension of the people who inhabit threatened places.

Building on the work of Schlosberg and others, a theory of biocultural justice emphasizes the diversity of epistemologies of maritime place, linking alternative knowledge systems (both traditional forms of ecological knowledge and scientific knowledge) with principles of sustaining ecological security. A theory and practice of biocultural justice should support relocation measures that sustain local and place-based food, water, and energy security, and protect the diverse epistemologies (reflected in language, knowledge systems, and traditions) that are essential to the maintenance of maritime identity. A theory of biocultural justice seeks to sustain the coevolutionary aspects of the unique social and ecological interactions of place-based peoples across time and space, even if a particular community is relocated. Three tributaries of biocultural justice are: the maintenance of the basic needs and functioning of maritime community structures and processes; the protection of maritime identities against the socio-ecological insecurities brought on by climate change; and the adoption of relocation measures that support the capabilities and functioning of living social and ecological marine systems across time and space.

Biocultural justice represents a shift from "shallow," anthropocentric theories of social justice to a "deep" practice of an ecologically based theory of justice. Schlosberg and others support a spatial expansion of the epistemology of justice horizontally into a broader ecological range of social issues and vertically into examinations of the global nature of injustices that are associated with food, energy, and water insecurity. This conceptual shift underscores the need to support a theory of justice that represents a deeper realm of human relationship with the more-than-human world where protecting marine ecosystem health and integrity are understood as essential principles and conditions that can support a practice of justice. A concerted effort at the international level is needed that can support a theory of blue justice. For those who receive the burden of the socio-ecological costs and risks from climate change, international agreements and conventions are needed that can encourage the protection of people who will need to be relocated in the future.

THE RETURN OF THE CANOE

We can return and restore the meaning of the canoe, which is something more than a life boat that only protects those with the wealth and power to climb on board (and prevent others from climbing on board). In chapter 1, I described the sense of Chumash inhabitation of south-central California that extends across the Santa Barbara Channel to the northern Channel Islands. The indigenous cultures of the Chumash embraced a bioregional sensibility that reflects an intimate relationship with islands, coasts, the Santa Ynez River, coastal watersheds, and the three transverse mountain ranges of the coastal province. The Chumash adapted to climate change that occurred during their inhabitation of south-central California. Major droughts occurred in California during the thousands of years of their inhabitation. One response included the use of canoes, which contributed to their survival. Trees were carved out into great canoes known as *tomol*s. The hull of each *tomol* was coated with the natural seeps of oil that washed up on shore. The *tomol* was used by Chumash to extend their reach to the Santa Cruz and Santa Rosa Islands, and for trading resources along the coast from Morro Bay to Malibu. At the bow of the canoe, an abalone inlay of swordfish or Elye'wun symbolized the voyage and the deep connection the people had with the sea. On their arrival on the island, a ceremony would mark the return to home place. Sage was burned. A seaweed dance or swordfish dance was performed celebrating the connection.

In the 1990s, the Barbarinos, the coastal band of the Chumash, restored the voyage across the channel by building the first *tomol* of the modern era. Crossing the channel in heavy seas, they traveled from the Santa Barbara harbor to Santa Cruz Island once again (figure 9.1). The abalone inlay on the bow of the *tomol* and a renewed dance were celebrated once again. Rebuilding the *tomol* marked the restoration of the maritime culture—a hopeful response to an endangered language, knowledge, and sensibility.

Across the Pacific Ocean, island peoples are restoring their connection to the sea by building ancient vessels that once carried people across

FIGURE 9.1 Paddle by Chumash of *tomol* to vakas off Malibu, California. (Image: M. V. McGinnis.)

Oceania. In the 1980s, native Hawaiians built a canoe called the *Hokulea* using old drawings of the ancient sailing vessels used by Polynesians to sail across the South Pacific. The *Hokulea* reflects a story and a diverse cultural heritage that created a message of a two-thousand-year-old traditional voyage across special places and the sea. It is a story that was almost lost to the colonization by Europeans of Pacific Island states. It is also a story of survival, rediscovery, and restoration of the traditional maritime ways, the language of the waves, winds, currents, marine life, and cosmos. Early Polynesian navigators sailed by the stars, following the path of great migrating species like the blue whale and the *honu* (sea turtle).

Another Pacific voyage began in 2013 and is referred to as the *Malama Honua*, Polynesian for "to care for our Earth." The voyage combines the canoes of the *Hokulea* and *Hikianalia*, which are currently sailing the world's ocean to highlight the diverse cultural and importance of "Island Earth" (figure 9.2). Across the world's ocean, the Pacific voyage represents a shared desire by island peoples to reconnect places and peoples, and to forge a global connection that can protect the values held by island peoples. In the voyage across the sea, the story of the canoe is

FIGURE 9.2 Sailing vakas. (Image: M. V. McGinnis.)

being taught and shared today. There is a sense of renewed hope in the face of an imminent crisis of both the biosphere and ethnosphere. With respect to the ethnosphere, the world is losing traditional ecological knowledge and language at an alarming rate. The rate of the loss of the world's diverse languages exceeds the loss of native species diversity.

With each returning Pacific canoe, the voyage is celebrated as origin and possibility, heritage and story. The return of the canoe is a reminder that we are on a blue island of finite resources, floating on an "invisible continent" in a sea of space. The canoe is also a metaphor for the places we inhabit and are connected to. The canoe lands on the coast, goes upriver, and travels deep into the watershed. The upper watershed is settled. We know that the boat we depend on is increasingly at risk, and we will need to collaborate and work together to keep it afloat. To build a canoe, there must be cooperation and a sense of community that is ultimately reflected in the voyage upriver, and to a particular place.

• • •

Thoughts from a voyage under stars in the great Pacific on a vaka (*wa'a* or canoe) . . .

The navigator, whale migration
In the wake of a tail, the vaka is a community of voices
Celebrate *malama honua*. Re-create
Impulsive oceanic gyre, atmosphere of cobalt, laughing children
Swirling flesh, fish in corals, in sunlight;

Sailing vessels, noble winds grow
Turtles dance in shallow seas; clouds roar
Remember to capture the wind with an open mind.

The moon is an owl's face gazing above you.
In the late afternoon, as I float on my back

The clouds are paint brush pink hue in a foreground that is the
     bluest salt smile
The sunset is about to burst open the night
Three-flipper turtle, come over, take a breath, take hold of the stars
Take hold of my heart, change me, skin to shell, illuminate my mind,
you have gone so far, without the left side, you are strongest, you
     have survived, swim circles in the clouds, blood sunset has
     opened, hide me
teach me the ways of the sea, shadows of you are below, the sun is
     your shell,
one more breath, change me, I will follow changed.
I swam to shore using only my right side.
I am your shadow.
Ancestors unite, teach the wisdom of the sea
Birth water, flow back to the sea
Link mountain and river and sea
Blue heart, saltwater veins
Learn to listen and hear the ocean is singing
We are all islanders, linked.
We re-member our origin
Deep history is more than human heritage or places we inhabit
The ocean's depth crosses many boundaries, as expressed in the gift
     of story

The ocean's story has gone into the wind. Moon rainbow in the
     night's light

CHAPTER 10

Restoring Place in the Theater of the Anthropocene

Precisely at the moment when we have overcome the earth
and become unearthly in our modes of dwelling, precisely
when we are on the verge of becoming cyborgs, we insist on
our kinship with the animate world. We suffer these days
from a new form of collective anxiety: species loneliness.

—Robert Harrison (1996, 428)

The declaration that "I am a fox" or that "you are a goose" is
the predication of an animal on a pronoun which is more or
less amorphous and helps to teach the art of metaphor. Just as
I say I may be foxy in strategy I can be a tree in my rootedness
or a rock in stolidity. Such multiple ritual assertions are a
kaleidoscope of successive, shared domains that define me
ever more precisely. My identity is not simply human as
opposed to animal. It is a series of nested categories.

—Paul Shepard (1996, 85)

My shovel digs into the soil as I try to remove the nonnative features of
the land. The red rash of poison oak is a reminder of my interaction
with the hillside's vegetation. Seeds, weeds, and sod cover my feet and
arms. I am thinking of Seamus Heaney's famous poem "Digging"
(1966). The nature poet moves language to a point of woven passages
and metaphors that invoke a sense of a primordial encounter with the
brilliant, often elusive landscape of his childhood. Heaney's life-affirm-
ing poem compares his father's digging of the soggy peat in the Irish
countryside to his own use of a pen. Heaney digs deep with rhythmic
order to plant ideas (instead of potatoes). A sense of beauty and

childhood memory is revealed through Heaney's language and use of metaphor. With my spade, I'm planting native monkey flower, Chinese houses, oak, pine, and Indian paintbrush. I am encouraged to move deeper into the landscape, and to remove the nonnative hemlock, eucalyptus, fennel, and yellow mustard each spring.

While I dig away at the European grasses on the hillside, I also recognize that the animals and plants that once were integral parts of this landscape are missing. The more intimate one becomes with a place's natural history, the more one recognizes what has been lost and forgotten. Most of us have firsthand experience of the impermanence in various aspects of our respective places. As if the workings of a clock have been removed, the key landmarks that once helped us return home are gone. Our home places have been transformed in a few generations, which have been preoccupied by growth and industrialism.

The sense of loss is a primary motivating force to deeply dig into our past to understand our relationships to one another and nature, and to move toward the greater community-based awareness that is needed to restore and preserve the unique facets of living in place. Restoring place is the focus of this chapter. Aldo Leopold (2001) recognized that restoration is a type of "intelligent tinkering" based on the cultivation of an intuitive, place-based knowledge and practice that goes well beyond a faith in the use of technology and science. Tinkering is the process of adapting and adjusting to one's surrounding with the hope of making repairs or improvements. Repairing, or better yet healing, requires a particular type of science and sensibility; it requires a shift in value orientation away from the pioneer mentality, utilitarian resource management, faith in managerial control, technological engineering, and global trade.

Intelligent tinkering is not heavy-handed, and restoration requires a particular pragmatic approach and respect, or what Gary Snyder refers to as the "etiquette of freedom." In *The Practice of the Wild*, Snyder (1990, 7) reminds us of the importance of wildness and wilderness: "It has always been part of basic human experience to live in a culture of wilderness. There has been no wilderness without some kind of human presence for several hundred thousand years. Nature is not a place to visit, it is home—and within that home territory there are more familiar and less familiar places." The freedom ushered in by the golden eternity of a springtime bloom, the acts of pollination, uninhibited reproduction, and recognition of the well-beaten paths taken by deer, bobcat, mountain lion, and birds within a watershed are based on an understanding and recognition of this culture of wilderness.

How do we start to unravel the impacts of the Anthropocene on wild culture and places? The challenge, in part, is a process of recovering an etiquette of freedom that respects the smell of the land, an awareness of the coming storm, the direction of the wind in the trees, and the behavior of animals. Restoration is a social enterprise that involves mindfulness of the intermingling of science and a renewed communal sensibility that is well-grounded in respect for the "biotic community."

NEGOTIATING NATIVITY

The idea of wilderness seems to be receding like the galaxies in deep space. Scientists who study the cosmos predict that the galaxies that are observable today will vanish from sight in the future. As the scale of the universe expands, future astrophysicists will not see these galaxies. Instead they will observe a vast darkness of space. Future observational astronomers will perceive the cosmos differently from today. Information about the future universe will change with expansion. Perhaps the future will be based on a sense of an isolated dark cosmology where we are left with only our own galaxy to behold. I wonder whether the scientists of the future will view wilderness in the same vein—a sense of nature absent of wildness as a more profound sense of our own species loneliness emerges.

The California flag is a symbol of a distant past that has faded with memory and cannot be recovered. Restoration defies technological fulfillment. On the California flag, the brown bear stands on wild bunch grass. There are few native grasslands in California; we've lost 99 percent of the native grasslands that once covered the valleys, foothills, and coastal plateaus of the state. The brown bear is stuffed, and remains behind the glass in our museums or in zoos. Early peoples followed the path of the bear to settle the continent. Before the introduction of beef and dairy cattle by Europeans, the great bear traveled to the grasslands of the central valleys of the San Joaquin and Salinas. During spring, the river valleys of the central spine of California were once full of wildflowers that attracted bees (with their honey) and bears. The presence of the bear was a sign of spring. Honey dripped from the bear's fur as it walked the central valley, along the creek or river corridor, to reach the sea and to feast on a beached gray whale with the California condor.

The subtle relationships and linkages between native honeybees, brown bears, and grasses are missing from California. Most of the 150 or so native tribes and their respective vernacular understandings of the ecosystems of California are also gone. In this Anthropocene age the plight of "nativity" is a critical one to consider as one begins the path toward restoration. Nativity implies an origin, a key reference point, and an ecology of interactions, relationships, and linkages that is based on the unique interdependent genealogy of our ancestors. This shared genealogy reflects a coevolutionary exchange between humanity and the more-than-human. Native people learned from these others, adapted, and evolved. For centuries, the native peoples of California set fire to hilltops and valleys to foster the germination of various plants that would be harvested and to increase prey species. The practice of fire ecology was essential to aboriginal culture across most Mediterranean climates. When Europeans arrived, their impacts on California's ecosystems were very different from the impacts that indigenous peoples had on the land and sea. Human evolution is a product of association with other animals, plants, and insects.

In this Anthropocene age, we cannot restore the genealogical foundation to ecosystems and the high degree of naturalness that once coexisted with human society. But we should take the initial step to restore our relationships to place for one simple reason: as in the past, cultural adaptation is based on an intimate relationship to nature. We need to become more aware and attentive to the changes that are occurring in nature so that we can evolve and adapt. Our industrial modes of activity threaten the remnant natural processes and essential ecological uniqueness of endangered nativity. We have also lost the intuitive knowledge that indigenous inhabitants have of natural places, and have become more vulnerable to the dramatic changes that are occurring to ecosystems. Our capacity and capability of adapting to the changes that are likely in the Anthropocene will be put to the test. Each generation of scientists perceives, studies, senses, and observes a different ecology than that which preceded them. Accordingly, it is increasingly difficult to understand the changes that occur in the natural processes of ecosystems. Changes in food and water supply, climate, and the essential life-producing character of ecosystems cannot be understood in purely scientific and technical terms. A deeper awareness of the character of ecosystem change requires that we dig deeper. As Snyder (1990, 24) so eloquently observes, "The wild requires that we learn the terrain, nod to all the plants and animals and birds, ford the streams and cross the ridges, and tell a good story when we get back home."

Knowledge of the local terrain also requires a new language. For instance, the term *biodiversity* encompasses the total number of genes, species, and ecosystems in a bioregion, and includes both native and nonnative animals, plants, and insects. Protecting biodiversity may not protect ecosystem health. Invasive species are the second leading cause of native species loss worldwide (Wilcove et al. 2000). For example, the Chinese mitten crab spreads disease to plants and mammals by way of introducing lung fluke, and generally degrades aquatic ecosystems, such as the San Francisco Bay (Cohen and Carlton, 1997). In some areas, biodiversity is increasing while native species diversity is lost. In marine areas, island biodiversity is particularly vulnerable to species invasions and climate change (Sax and Gaines, 2008). The biodiversity of California's Channel Islands has increased by 44 percent (Sax and Gaines, 2008). Despite efforts in some regions to control new introductions, more exotic plant species will be *added* to islands and other coastal areas of California over the next century.

The remnant nativity of a region is worth preserving, for without it we lose wildness. It is not just preserving a particular species or group of species that matters. We need to focus on protecting the relationships and connections between species—their habitat and prey species. This chapter's focus is on the need to recover the human relationship to place. One problem is that as we denature nature we have less of a language and knowledge of place to draw from. Accordingly, restoration should be thought of as a movement forward to a recovered sense of place and community. A particular culture defines the landscape—it may be wild, tamed, indigenous, violent—but these terms become less meaningful in a context and industrialized culture that has lost its way home.

THE POWER OF MIMESIS

We are not mere passive observers of nature. We are not just members of an audience watching the spectacle of nature unfold on a stage. If we think ecologically, there is no "outside." Shepard (1996, 86) observes:

> Personal identity is not so much a matter of disentangling the self or "the human" from nature as it is a farrago of selected correspondences in which aspects of the self are projected into the dense, external world where they are discovered among a variety of animals who are both similar and different from us. Aspects of the animal are then reintrojected into our psyches by a wonderful chemistry of imitation. When we observe this unlikely agency at a distance, animals seem like mediators, appearing in music, story, song, narration, dance, and mime as participants in the narrative.

Restoration should be understood first and foremost as a form of *cultural mimesis*. Cultural mimesis is the human capacity to represent, ape, or imitate the natural world and other human activities. As a child learns to eat by watching its mother and father, so we can learn to restore our relationship to place by being more attentive to it. Mimesis is a process of culturally and imaginatively constructing a relationship to nature and others. It is in the spirit of imitation and mimesis that our denatured society can retrieve a sense of place. Restoration is derived from knowledge and partnership within a community, and the reciprocal norms that govern community (McGinnis 1999b). Hands-on ecological restoration can be one means to restore the connection between the mind, body, and flesh of the world we inhabit.

Michael Taussig (1993, 106) uses mimesis in verb form as "to mimeticise," which he understands as a mechanism of psycho-physical assimilation to a mode of being initially encountered "outside" the self. Taussig's influential *Mimesis and Alterity: A Particular History of the Senses* characterizes mimesis as "the capacity to Other" (19), which "tak[es] us bodily into alterity" (40) and sustains ritual enactments and totemic practices. In this sense, mimesis can be one way to assimilate human beings and place, a way of crossing over to one's natural terrain. Mimesis is an important part of philosophy that begins with the early work of Plato and Aristotle. It is a concept used in theater, science, politics, anthropology, and ethics by contemporary scholars. The value of mimesis is described again and again across the centuries, sometimes under conceptualizations linked to the Greek philosophy of mimesis—like *imitatio* in Latin, *imitation* in English and French, and *Nachahmung* in German—but also in terms like *sympathy, contagion, identification, empathy,* and *intersubjectivity.*

Like Taussig, I am most interested in thinking about mimesis in terms of the relationship between human society and nature, and the role of mimesis in supporting a type of redemptive journey toward a recovery of place. "Once the mimetic faculty has sprung into being, a terrifically ambiguous power is established; there is born the power to represent the world, yet that same power is a power to falsify, mask, and pose" (43).

The concept of mimesis is part of Greek drama and Celtic action. In Celtic action and artistic expression, mimesis reveals the material and spiritual forms of nature. In antiquity, dramatic acts were attempts to represent the physical world through art and language. According to Aristotle, tragic poetry and mimetic art disclose a systematic and logical unity in the universe. The artistic representation of the natural world in tragedy serves as one means to unite the human condition with the natural world, to allow for the mediation of the real and ideal (Plato), to support the synergy between the particular and the universal (Aristotle), and to serve the theoretical articulation of universal forms such as the Good, Truth, and Beauty. The representation of these universal forms symbolizes the embodiment of the classical project—especially for Plato, the first Western philosopher of mimesis.

In another example, the craft of Pueblo pottery embodies the tribal beliefs and orientations that have sustained the Pueblo culture. The pottery includes an array of human "clay people," animal figures, and components of the landscape. The clay and minerals that make up pottery come from the earth. The clay is delicately formed and prepared by the hands and emotions of the craftsman. Clay becomes pottery (becomes animate) without the use of a machine. The earthly clay and the tribal story are united. Each finished pot represents the fusion of the earth, an elder's story, and the image. This image is the mimesis of a story told by the storyteller (the figurine). The figurine is more than a commodity of industrial culture, or an artistic form, or an aesthetic representation of culture. It is an embodiment of cultural memory, which preserves the stories of a particular place and custom, and the dreams and inspirations of the craftsman.

MIMESIS AND WORLDVIEW

There are several restoration cultures and discourses, which include quite different agendas, perspectives, and worldviews that shape mimetic activities. Worldviews are shared experiences which shape our sense of self (our identity), our capacity to cooperate, and our personal choices. A worldview is a picture of the way things in sheer actuality are, our shared concept of nature, of self, of society. It contains our most comprehensive ideas of order. A worldview is made emotionally acceptable by being presented as an image of an actual state of affairs of which a way of life is an authentic expression.

Human beings develop various worldviews to bring order to human existence. Each worldview is a bundle of values and principles which shape identity, community, and culture. Some worldviews lead to more destructive human behavior than others. During the 1920s and 1930s, Walter Benjamin, a European intellectual, examined the materialistic underpinning of contemporary art and literature, and the displacement of urban and industrial society. In *The Work of Art in the Age of Mechanical Reproduction* (1977 [1935]), Benjamin repudiates the illusions and reproductions of mechanical society. For Benjamin, modern technology and the machine could reconfigure cultural heritage and tradition, bringing about the destruction of the authenticity of origin—"even the most perfect reproduction of a work of art is lacking in one element: its presence in space and time, its unique existence at the place where it happened to be" (222). With mechanical reproducibility, we lose the sense of origin and reference.

Theodor Adorno (1984) also described the impact of mimesis in an era that he referred to as "the culture industry." The transformation of a culture of wilderness to a culture industry marks the rise of the mechanical sensibility and the influence of global capitalism. Adorno was part of an intellectual movement known as the Frankfurt School, which valued mimesis as a potential way to resist the power of capitalistic culture. Adorno mourned the loss or repression of a primal capacity to mime "nature" within a culture of instrumental rationality. In *Aesthetic Theory*, Adorno called for mimetic activity that would "endeavor to recover the bliss of a world that is gone." His particular focus was on artistic creation. For Adorno, art should break from the culture industry to speak for the beauty that is found in nature: "The beautiful in nature flashes out, only to disappear immediately when one tries to pin it down. Art does not imitate nature . . . but rather natural beauty as such" (107).

In their philosophies on mimesis, Adorno and Benjamin embraced an aesthetic view of nature—nature defies human language and understanding. In this sense, nature is detached from the human experience. This view of nature, however, fails to reflect the fact that we are part of the natural world. Paul Shepard (1996) reminds us that myth, folklore, story, religion, art, law, and norms evolve with the mimetic act. Unlike the aesthetic theories of Adorno and Benjamin, Shepard describes several cases of the celebration of nature, the embodiment of nature, and the intimate and interdependent relationships humans have with the natural world. Adorno and Benjamin maintain a view of an aesthetic nature to behold, while the human ecology of Shepard shows us that humans participate in and relate to nature in many ways—culture and nature evolved as a unified whole. The human being is not autonomous or separate from the natural world.

MIMESIS AND THE ANTHROPOCENE

There are forms of mimesis that contribute to the degradation and falsification of nature. Technocentrism is a core belief structure that supports a choice to control and manage nature (O'Riordan 1995). As noted in chapter 3, Riley Dunlap and colleagues (2000) consider technocentrism the dominant social paradigm and environmental worldview in industrial societies. To the detriment of a relational and communal mode of being, technocentric mimicry supports industrial modes of mechanical simulation (McGinnis 1999a, 1999b). My focus here is on the values that support the mechanical sensibility and how it changes our relationship to nature and society. Industrial and mechanistic values have existed in Western culture for more than a century. My focus is on the mimesis of industrial values that supports globalization and the mass production and consumption of goods, things, and materials. In industrial mimesis, the global reconfiguration of communal economies moves the member of a community and place into the role of a spectator, an observer of nature, a consumer-producer of things. The new global consumer stands outside in an "environment" while nature becomes an object of inquiry, a resource to be taken for granted, to behold, to be traded and ordered-for-use. The mimesis of industrialism across the globe is not tied to a particular place, locality, or bioregion.

An example of the technocentric worldview was given in chapter 3. More than a thousand fish hatcheries produce approximately 80 percent of the "salmon" in the Columbia River. Fish hatcheries have been used in the Columbia River basin for more than a century to enhance the economic and cultural values of salmon. The synthetic salmon cannot equal a wild salmon because the histories and the ecologies of the two "animals" are dissimilar. The machine and the image of the salmon are developed simultaneously. The engineered salmon is a new life form, but it is incapable of manifesting all of the unique features of the original.

Another example is the "organic machines" associated with offshore oil development described in chapter 5, and public policies that support "artificial reefs." Artificial reefs cannot serve the function of naturally occurring marine systems.

My concern is that industrial mimesis (with its engineered and synthetic images) is a sign of a new psychic structure of mimesis—a mimesis that perpetuates a disconnection of culture from place. We live in a mechanical cage that no longer respects the psychophysical unity of a community of living organisms. Industrial mimesis is rapidly becoming the prevailing mode of constructing the world. By the process of industrial mimesis, the salmon is replaced by its double, its copy, the human-engineered salmon. The illusion of the salmon restored by the machine and engineer is hyper-realized, while the real salmon is denuded and denatured. The possibility of the disappearance of the "salmon" and the "river" is the final product of industrial mimesis.

Jean Baudrillard spent the last several years of his life writing on the paradox of mimesis in modern society. For Baudrillard, mimesis is a mode of simulation that is far removed from an original reference to nature. In *Simulacra and Simulation*, Baudrillard (1994, i) writes: "Today abstraction is no longer that of a map, the double, the Mirror, or the concept. Simulation is no longer that of a territory, a referential being, or a substance. It is the generation by models of a real without origin or reality: a hyper real. The territory no longer precedes the map, nor does it survive it." Baudrillard maintains that we cannot rely on our museums, antiques, technology, engineering, embalming, cloning, models, or other facets of simulation that are far removed from nature. Indeed, restoring our relationship to the community and nature is made even more difficult by the diminishment in the diversity of the circle of animals and plants that were once part of our lives. With the diminishment of life, our field of being is reduced. Most of us inhabit mechanized and citified environments that are far removed from an original earthly home, and we find ourselves misplaced in a cosmopolitan hyper-reality. Our shared ecological identities have been taken, broken, transformed, and replaced by our love of machines and the work of our doubles.

This is not to imply that nature is not wild or that some semblance of natural processes no longer exists. But the reference or origin of nature changes in technocentric mimesis. Restoring primary nature to its substantive form, its original ecological substance, its referential being is doomed to fail in an industrialized and hybridized context. Industrialized environments are now recognized as primary nature, while our specific bioregion is perceived as secondary nature and labeled "the environment." Since there are few signs of "real" and original nature left, the fear is that restoration can quickly become yet another product of a "hyper-real" industrial age. The sounds of our machines simulate the sounds of spring and birds, while the spring bird migration diminishes. The sounds of a river's spring runoff are missing, while the sounds of hydropower development fill the air. There is a profound sense of hyper-reality to our mechanical existence. Our ability to distinguish between real (nature) and illusion (artifice) is blurred. The next step in this contemporary period of B-movie science fiction and mechanical reproduction is the Human Genome Project, the ultimate act of replicating denatured humanity.

In technocentric mimesis, society's ability to mime nature no longer flows directly from daily observation. This is an important point because the restoration of "nature" or the "renaturalization of nature" is essentially a mediated activity; the act of restoration is a product of social negotiation that includes science, values, and epistemologies. In a scientifically oriented society like that currently prevailing in the West, we tend to think about an activity such as ecological restoration in scientific and technological terms. Scientific investigation, however, is conditioned by cultural values.

VALUES AND SCIENCE IN RESTORATION

The mimetic faculty is a physiological mechanism that serves socialization. As a type of mimesis, restoration is a form of creating secondary nature. Values determine the process of restoration as a mimetic act. It is important to note that in making this assumption I reject the belief that a fact–value distinction should be made. Rather, I propose that facts and values are integrated parts of restoration activity and practice.

One study, conducted by my colleague John Woolley and me, focused on the ways in which values and perceptions of science affect the way scientific information is used in restoration of river systems (McGinnis and Woolley 1997). A characterization of our work was given earlier, in chapter 3. We investigated the range of values and beliefs held by restorationists involved in three large-scale river projects on the Yakima River in central Washington, in the Upper Sacramento River basin, and on the Santa Ynez River in California. On the Upper Sacramento River, defined as the portion of the river and its tributaries between the Feather River and Keswick Dams, a distance of 358 river kilometers (222 miles), salmon and riparian habitat have declined over the past three decades. A hundred and fifty years ago, the Sacramento River was bordered by nearly 200,000 hectares (500,000 acres) of riparian forest, with riparian vegetation extending six to eight kilometers wide. Development in this area, however, has led to a 95-percent reduction in riparian habitat since the 1960s. The Yakima (named after the Yakama Indian Nation) drains much of south-central Washington before emptying into the Columbia River. The upper basin is National Forest land, while the Yakama Indian Reservation occupies much of the basin west of the river and below the city of Yakima. There are several restoration groups and organizations associated with these three river systems.

To characterize the values and beliefs of those involved in restoration, we prepared a mail survey that included sixty-five statements related to restoration efforts, and asked recipients to indicate their response to each of these on a five-point scale from 1 for "strongly agree" to 5 for "strongly disagree." Using lists of watershed groups active in these three river restoration efforts, we sent surveys to a random sample of 421 persons, including preservationists, landowners, members of indigenous communities, property rights advocates, scientists, and personnel from resource agencies. The surveys were sent not only to respondents who were actively involved in restoration but also to less active, but informed, members of the community. The respondents represent a culturally diverse population of citizens involved in restoration planning. One hundred forty-nine (35 percent) responded to the first round of surveys.

What we found was surprising. While respondents differed in their reactions to many statements, in general it was not questions of values and beliefs but questions related to science and its application to restoration that elicited the widest range of responses, or, to put it another way, the highest level of disagreement. What this suggests is that people may be more likely to agree on issues of value than on scientific and technical issues.

Our study supports the premise that restoration science and practice exist in a larger context of values, or what we referred to as a *restoration concourse*. A concourse is a variety of existing opinions bearing on the issues associated with a particular activity—in this case restoration consists of shared perceptions and values. Each discourse rests on a shared set of assumptions regarding the purpose and values of a restoration effort, but commonly, as our survey results suggest, a restoration project will include people committed to more than one discourse, some of which may be incompatible (McGinnis and Woolley 1997). If these discourses are recognized and discussed, we believe that the development of restoration strategies and plans will be more successful both ecologically and culturally.

The first step in doing this is to identify the discourses that emerge as a project takes shape. While restorationists will ordinarily do this intuitively, we used factor analysis of the results of a second type of survey instrument to identify the discourses being carried on in association with the three river restoration projects that we studied. In this way, we identified four more or less discrete discourses. Two of these were shared by only a few of our respondents, and would probably have little influence on the outcome of a project. The other two represent discourses that are shared by a larger number of respondents and are therefore more likely to influence the outcome of a restoration project. We should note that we also identified several additional discourses and sub-discourses, two of which are anti-restoration and anti-technology discourses. Here again we found that, while certain technical and scientific issues are centers of disagreement, neither of the two prevailing discourses is strongly organized around issues of scientific knowledge or technology.

The first prevailing discourse is a normatively based restoration discourse that reflects the assumption that restoration is both factually necessary and ethically mandated. We called this the "restorationist" discourse. Those sharing it take it for granted that scientific uncertainty is real, but that this should not be regarded as an obstacle to restoration efforts. They place a high value on restoration, but recognize its limitations and are generally pragmatic about it. Thus, they place a high value on the natural landscape, and share a faith in the possibility of restoration, but disagree strongly with the statement that "the environment can be repaired just as a mechanic can repair a machine."

The second prevailing discourse is also prescriptive, but places far greater emphasis on private-property rights and local control of projects. Respondents sharing this discourse place a high value on property rights and local control, view the claims of both restorationists and scientists with skepticism, and emphasize that restoration necessarily involves trade-offs with other important cultural values. They are especially concerned about the conflict between restoration and property rights, and typically endorse the statement that "environmentalists have overstated the need to restore the environment." Hence, they believe that "restoration concerns are as much philosophical as they are technical," and accept the view, rejected by the restorationists, that "people can manage, manipulate, and repair the environment just as a mechanic can repair a machine."

PATHS OF RESTORATION

How modern society has come to relate to place, science, technology, and nature influences the discourse of restoration. Whether we are talking about conserving, preserving, or restoring nature, our treatment of the natural world—expressed as unique, unexpendable, but interrelated places—is essential. Individual and cultural identities extended to include the lives of other species can serve to unite a culture with nature and place. This ecology of shared identity is a mirror of the multiplicity of place—a place that includes a circle of animals and habitats. In perhaps the most comprehensive inventory of how and why humanity relates to place, Shepard (1996) shows how a culture's ability to adapt and endure is dependent on context, modes of understanding and organizing, natural history, beliefs, and values.

As the bioregions which are our field of being are reduced by the extinction of species and the pollution of soil, air, and water, humans suffer from a condition of diminished health and perception. As the well of natural provision dries up, we lose our inclination to treat each other generously. Our ability to analyze our situation and our relationships with place is reduced. We find it difficult even to think clearly about our condition. This is the condition within which the restorationist works: we are disabled creatures dislocated in a wounded landscape. How we organize to deal with ecological crises will determine our shared fate. Species loneliness in a wounded landscape moves us to want to restore our relationship with place and others, or to put it another way, modern humanity yearns to re-establish and restore an ecology of shared identity. Rather than understanding the world through a relationship with earthly entities, modernity favors the human ability to experience nature as a quality (or quantity) that springs from scientific, technological, bureaucratic, and economic understanding. Human beings remain isolated actors in an earthly cage: the world is technologically divided, scientifically categorized and manipulated, and perceived as absent of spiritual and intrinsic worth.

The discourses of restoration are part of a negotiating process that shapes and influences our relationship with nature and community, and these discourses also shape the mimetic act. Based on values and different forms of knowledge, there are many paths that restoration can take. There are two primary paths of restoration: *isolate* and *bioregional* (McGinnis, House, and Jordan 1999). Isolate restoration is based on the values of modern technology which support the separation of place from culture, while bioregional restoration supports the reintegration of culture with place. When a community engages the challenge of restoring its ecosystem functions, it is embarking on a self-educating and culturally transformative path.

The Isolate Restorationist

The isolate restorationist finds it difficult to grapple with the continuum of human interdependence with a healthy, self-sustaining bioregion and community. Many individual scientists recognize that the health of human communities and the health of bioregions are coterminous, but the tools of their trade—their reductionist rigor—often prevent them from engaging the messy interpenetration of humans and places without violating their own discipline. Rather than engaging the interpenetration of human communities and the landscapes they inhabit, the scientific "eye" embedded in the body of isolate restoration efforts can serve to reinforce the subject/object separation of place and human community. Donald Worster, in his definitive study of the development of ecological thinking, *Nature's Economy* (1979, 289), illuminates this trend in the discipline of academic ecology in the latter half of the twentieth century: "At the very moment [ca. 1949] he [Aldo Leopold] embraced it as the way out of the narrow economic attitude toward nature, ecology was moving in the other direction, toward its own niche in the modern technological society. It was preparing to turn abstract, mathematical, and reductive."

Despite our technological intentions, place continues to influence human activities and cultures in countless ways—altering our habits, cities, cuisine, language, values, and expectations. Moreover, the loss of a species or a mountain is not merely a failed experiment. The loss of a species represents the diminishment of our perceptual field of vision. We should attempt to restore the human relationships and shared perceptions that define a community of place. A mimetic relationship that blurs the boundaries between the subject and the object, and the division between place and society, is needed. In such a relationship, "there will no longer be a humanity, or a nature, but a continuum of connection that is the primal asking force" (Rothenberg 1996, 265).

Human beings can care about, fear, defer to, and separate themselves from natural entities. A wild place or nature reserve can be recognized as "holy wilderness," a "recreational facility," a "natural resource," a "scientific experiment," a "zoo," or something to be avoided or conquered. Each of these reflects a particular human–nature relationship. The values and perceptions are also representations of particular worldviews. In shallow ecology, nature is bureaucratically managed, subjugated, structured, enframed, viewed as a means to an end, and ordered-for-use as a "natural resource." In public life and on public lands, a mountain understood as lacking any reasoning capacity is denied moral consideration, bombed, its soil leached with cyanide, its "timber" cut, mined for its gold. The grizzly bear, lacking the capacity to be reasonable or self-conscious, is considered "spoiled," bad, feared, and managed out of existence. This is the essence of the shallow ecological movement.

Bioregional Restoration

There exists a deeper worldview that shapes an alternative ecological relationship. This deeper realm of the restorationist–nature relationship is the issue I now turn to. There is a second path for ecological restoration: *bioregional* restoration. The goal of bioregional restoration is to re-immerse the practices of human community within the bioregions that provide their material support, as well as the direct relationships to the more-than-human world on which the full range of human experience depends. Bioregional restoration is a performative, community-based activity based on social learning and cooperation. If local communities are left out of the process of restoring the landscape and place, then restoration is not bioregional. Bioregional restoration can be a therapeutic strategy to expose ourselves viscerally to local ecosystem processes, to foster identification with other life forms, and to *rebuild* community within place, as the insights and local information that emerge from restoration activities affect the cultural and economic practices of the human population. Some differences between the values of isolate and bioregional restoration are described in table 10.1. Note that the isolate restoration path and the bioregional restoration path are not mutually exclusive. If the discipline of restoration ecology had not risen independently, attainment of re-inhabitation would have required its invention. Any restoration effort that attempts to engage the industrialized landscape is dependent on the insights of ecological science. The practitioners of bioregional restoration, grounded in particular places, rarely have the luxury of isolating the human presence from the more-than-human world; neither can they afford to eschew the powerful tools that science provides.

Promoting shared living place is not the goal of isolate restoration. Scientific evidence supports the reintroduction of the wolf to the Greater Yellowstone Ecosystem. The wolf is viewed as an important predator in that ecosystem. Restoration of the wolf to Yellowstone, however, requires more than the reintroduction of the wolf to the system. The sensual, sacramental, spiritual, and ecological values that are endemic to a healthy wolf population and the wolf's place in a diverse Yellowstone bioregion should be respected, cared for, and ultimately restored. Relationship building is not a primary concern for the isolate restorationist. In contrast to isolate ecological restoration, bioregional restoration requires an alternative relationship with these Others.

The bioregionalist begins from a different set of motivations and constraints from those of the isolate ecologist. Place may be scientifically

TABLE 10.1   A COMPARISON OF RESTORATION PATHS

	Bioregional restoration	Isolate restoration
Function	Communion	Observation
Social dimension	Community practice	Management based on expertise
	High degree of interpenetration between cultures and place	"Nature" studied in isolation from human influence
Technology	Focus on locally appropriate technology	Coexistence with industrial production
Science	Experimental	Data-based
Activity	Preservation	Replacement
	Cooperation	Domination

defined by its geomorphic, ecological, and hydrological characteristics. For the bioregionalist, the scope is expanded to include the degree to which local communities enfold themselves within the constraints and opportunities of particular places. Human cultural definition from within the bioregion plays as large a role in the definition of place as do the more quantifiable nonhuman aspects identified by isolate science. Bioregional restoration is a *practice* performed by a community that extends its identity to biospheric life as manifested by particular places; a human community which begins to define itself through its continuity with and immersion in ecological systems.

Bioregional restoration is, first and foremost, a service we offer to nature and to each other. And at the same time, by giving us work to do in the landscape, it satisfies the first requirement of membership in the community. Bioregional restoration is not only a service we offer to nature out of a sense of duty, it is also in many instances a gift, offered freely out of love and affection for a place. As poet and philosopher Frederick Turner points out in *Beauty: The Value of Values* (1991), an exchange of gifts is always to some extent problematic because we can never be sure that what we give is commensurate in value with what we take or have been given. This is true in the case of our relationship with the rest of nature: since nature gives us all we have, including life itself, how can we ever repay it?

We tend to think that relationships and communion are easy, natural, and perhaps free of tension and uncertainty. This view, however, is clearly a peculiarity of our own modern civilization. Premodern cultures continue to view community and nature differently—not as easy but as perilous and uncertain, and not as "natural" but as an achievement of the human community acting together to confront this uncertainty and find ways of coping with it. This is clearly evident in the ritual technologies premodern cultures have developed to perform the work of community-making, and, as described earlier, is carefully explored by Shepard (1996) in his portrayal of the cultural mimesis of the natural world. Rituals of mimesis and initiation, by which a child achieves membership in the human community, often involve humiliation, ritual death, and the mutilation of the body. The rite of communion itself, by which the human community negotiates its relationship with the more-than-human world, begins in the act of killing, and represents an attempt to come to terms with the fact that life depends utterly on death. Destruction of "nature" is very much a part of restoration activity—as exemplified by the removal of "exotic" species from a riparian area. Hence, destruction and construction are part of restoration work.

The key role of bioregional restoration is the building of a human community the self-definition of which is extended to include the larger biotic community. Place-based ecological restoration can provide the shared experience, knowledge, and ritual necessary to such an undertaking. This is not a solitary experience but rather lends itself to group effort and even to celebration and festival. Bioregional restoration must not only deal with the historical degradation of ecological processes due to human practices but with the artificial boundaries that separate the inhabitant from his or her own local habitat, as well as the variety of values represented by human residents.

However much our thinking minds may spin off in the direction of technological invention and the comforts of a controlled environment, our senses remain immersed in a bioregion which is not entirely of our construction or invention. The bioregion is the source of our deepest pleasures. Our current alienation from these processes may not be as profound as we sometimes fear. The initiation of community-based ecological restoration projects is a powerful context in which to put the combined tools of science and bioregional sensibility to work in the service of personal and community transformation—to begin the process of reorientation of inhabitant to habitat, of community to place. Such projects inevitably lead their practitioners to an ever-deepening collective experience of the processes of place, and move them to confront the barriers that separate them from it. Bioregional restoration offers the opportunity to imagine ourselves back into our place-worlds while maintaining the evolutionary continuity of the communities of flora and fauna which define the particularities of each place.

REWILDING IMAGINATION: THE ROLE OF
ART AND THEATER

The imagination plays a critical role in the path toward bioregional restoration. Bioregional restoration requires more than replanting native plants along a creek's bank, digging and removal of invasive species, or restoring a river's flow to bring back wild salmon. Bioregional restoration also includes storytelling, theater, art, and ceremony that strengthen the connection between people with place.

One of the greatest examples of the affirmation of life and the regeneration of the wildness within is found in the black magic primitivism, poetry, essays, travels, drawings, dramatic art, and "theater of cruelty" performed by Antonin Artaud between 1920 and 1945. In many ways, Artaud was a rare shaman dancing in a dislocated era. His goal was to "slough off his European skin," to feel the red earth, to be absorbed by the sun.

> Thinking means something more to me than not being completely dead. It means being in touch with oneself at every moment; it means not ceasing for a single moment to feel oneself in one's inmost being, in the unformulated mass of one's life, in the substance of one's reality; it means not feeling in oneself an enormous hole, a crucial absence; it means always feeling one's thought equal to one's thought, however inadequate the form one is able to give it. (Artaud 1976, 70)

In 1933, Artaud delivered his essay "The Theater and the Plague" at the Sorbonne. During the lecture, he abandoned the text and began to "act out" the plague itself. "Lucifer and his beings have got me!" he later wrote. "They always want to hear about; they want to hear an objective conference on 'Theater and the Plague,' and I want to give them the experience itself, the plague itself, so they will be terrified, and awaken. I want to awaken them. They do not realize they are dead!" (quoted in Hayman 1977, 89–90). Taken by the wildness within, the death of Artaud, the "actor," coincided with the pain and agony of the reenactment/restoration of the terror of the plague.

Artaud's Theater of Cruelty envisioned a newly reborn social space, reunited with the life forces of an animate world. The Theater of Cruelty was inspired by prehistoric spirit possession and purification rituals. As with primal mimesis, Artaud's theater is intimate and sensual. It is hard to absorb and take in. The shocking character of this theater led to its demise. In 1935, Artaud's experiment with the Theater of Cruelty closed after only two weeks. Despite his performative abilities, Artaud was an intellectual without a community—a fate that took a tragic toll on his well-being.

The life of Artaud and the voyage of his mind have faded from our collective memory, but the extended reaches of his mind and work have stimulated contemporary theater, art, and philosophy. Artaud's wild imagination was celebrated by individualists like Baudelaire and other romantics who were threatened by "mechanical" objectivity, "group-think," and the impacts on the arts and society of industrial progress. Baudelaire and the French avant-garde valued originality that was not a product of mimesis or imitation but the fruit of the artist's own nature (Daston 1998).

The public events or "happenings" of the mid-twentieth century were also inspired by wild imagination. Happenings were tribalistic and spontaneous artistic events held in the late 1950s in New York, Los Angeles, Chicago, and other cities. The term is derived from Allen Kaprow's 1959 performances, *18 Happenings in 6 Parts*. Kaprow proposed that the artist should be an "un-artist" who attempts to transform the global arena rather than produce marketable objects. Another early art activist was the composer, poet, and teacher John Cage. Cage taught that art could evoke the random character of nature and culture. These "happenings" were the seeds of the "performance art" that followed. Performance was more than an act, it became an event.

As an event, the goal of performative theater is to get things done. Art historian Kristine Stiles (1996, 721) writes: "The important thing these various creative attempts have in common is that rather than seeking integration into the industry they seek to disrupt it. These experiments in liberation theater are in open conflict with the capitalistic environment." These forms of artistic expression are explicitly political. The theater and images reflect the human predicament, the terror and violence of their respective contexts, the brutality of war, a rejection of existentialism and materialistic society.

These key aspects of transformative theater and art in a dislocated era are reminiscent of the "epic theater" proposed by poet and playwright Bertolt Brecht. Between 1918 and 1956, Brecht developed a theory of epic theater that incited social "interaction"—no division between the "aura" of the audience or performer existed. Epic theater was a political expression that did not reproduce social conditions but rather revealed them. Brecht's particular concerns were related to class-based struggles that existed in Europe. He did not hide his Marxist views. Brecht proposed that theater, art, and literature should change our social and economic relationships with one another. Epic theater breaks from the linear development and evolutionary determinism of bourgeois theater. He believed that epic theater and the "learning play" (*Lehrstücke*) could change social and environment conditions.

Epic theater, however, is hardly an ecologically oriented art form. "Faced with the river," Brecht wrote, "it consists in regulating the river . . . faced with society, in turning society upside down" (quoted in Bartram and Waine 1982, 39). Epic theater's primary focus was on social and political transformation while maintaining the anthropocentric character of society. Nature remained to be managed and controlled. Unlike the guerrilla theater and street happenings of the 1960s and 1970s, Brecht's vision of the epic theater was based on the industrial value of mastering nature.

My view is that art and theater can restore society only if they also attend to nature, and the human relationship with nature and community. A "restorative theater" enhances the individual's capacity to experience nature and enriches the communal sensibilities with others. Restoration should be more than an art form that serves "functional beauty," objectivism, individualism, self-indulgence, or modern aesthetics. Good restorative art and theater serve the community, not the ego, and should take on an ecology of their own. Their ability to foster community building and partnership is the true test of their lasting force. Bioregional restoration inspires a form of collective learning.

One example of the use of theater as a restorative tool is Human Nature, a theatrical group which was created in 1985 to "re-examine the relationship between human nature and the rest of the world." David Simpson, a founding member of Human Nature, states that "theater can show the larger context, where biology and society inevitably meet, where the real choices that affect the future are made. If it can do so with humor and grace, the result becomes more than just educational—it can be a force for transformation" (quoted in McGinnis 1999b). In the early 1990s, one of their plays, *Queen Salmon*, toured the Pacific Northwest and quickly became an important contribution to a number of bioregional efforts to restore salmon.

TOWARD AN ECOLOGICAL MYTHOLOGY

Mythmaking and storytelling are important aspects of mimetic activity that can unite nature with community. Anthropologist Claude Lévi-Strauss (1978, 4) suggests that "what we discover is the simple opposition between mythology [the unreal] and history [fact] . . . is not at all a clear-cut one, that there is an intermediary level." What we have accepted as real or fact may merely be the dominant myth that we have decided to live by. Although it appears to be real, it may only be for the time being. Here the word *myth* shifts its meaning, from myth as illusion, to be contrasted with fact, to myth as inescapable world view, by which each of us must live. Because we have no non-mythological access to facts, that is, no facts that come to us unfiltered by our worldviews, the problem becomes a judicious choice of the best myth. In ecological affairs, however, this choice is not simply a personal one; rather, it is a collective choice, since the myth we choose determines which kind of social relationships influence our access to nature. Feedback and feed-forward loops are established: our social and ecological relationships are the results of our choices, and they dictate what choices we can make.

An ecocentric myth, or what I call an eco-myth, offers a principal medium for the politicization of a more intimate human relationship with place. Eco-myth can inspire a type of mimesis that is transpersonal and is exemplified by the myth of Aldo Leopold's land ethic. In Leopold's (2001) story of his journey, he kills a wolf and witnesses the green light in the wolf's eyes fade. Leopold deepens his understanding of the wolf and its role in the Arizona ecosystem. Although he has been taught as a professional to practice predator control to increase deer populations, he also notices the range of impacts of increasing deer populations on the countryside. Leopold's land ethic is in opposition to the myth of the taming of the continent. Leopold's eco-myth and recognition of the biotic community represent the integration of science and sensibility, and contribute to what he refers to as "intelligent tinkering" or the practice of restoration of place. Leopold's family purchases a small plot of land in Wisconsin that has been degraded. During the summer months, he and his family plant native trees. It is possible to visualize a kind of place-based practice where moral considerations are not repressed or kept apart but are systematically commingled with scientific investigation, where moral considerations need no longer be smuggled in surreptitiously, nor expressed unconsciously.

If there were such a science, in which moral considerations were juxtaposed with scientific inquiry, the positivism of the technocentric worldview could be rethought and questioned, making possible a new form of collective action through the power of an ecological myth. The division between fact and value, or appearance and reality, is not as clear as is frequently supposed. History can be understood as a continuation of myth. We live by the myths that we choose. Nature is experienced in terms of the myths that we collectively decide upon. Myths can become our reality. A new (old) ecological myth is one means of uniting the culture with place.

We may believe that an eco-myth in actual practice remains distant, vague, dreamy, aspirational, and impossible to live. We need to find a place and dig in. Place-based restoration can inspire a new eco-myth. Indeed, this is exemplified by Leopold's story. Leopold's identity shifted in meaning to encompass the biotic system. We have many eco-myths to draw from: storytelling, metaphysical speculation, symbolic speech, and political action are fundamental to myth development. The story of Leopold's journey is but one example of how an eco-myth is possible. A creative, imaginative dialogue, one that is relational in some deeper sense, can provide the needed inspiration toward a collective experience with place that can help us go beyond a myth grounded in technocentrism and embrace instead a myth in which we share in and unite behind nature's ever-changing character.

. . .

Why does silence fall across the night's sky, I wondered.
I stand still listening to a bird, the last bird of this day, call out
    before the golden eternity of this sunset
And the frogs in their soon-to-be orgy are beginning to croak
Insist on your own freedom, I said to them, throw the scripture
    away
Let's call the frog a frog, said Coyote, as I lay awake along the
    river's bank
The gold sun fell and I settled in. All is well.
Crickets and frogs go silent in the middle of night while I lie
    awake
In the river there is silence. Did I create this river? Yes, because it is
    what everything is.
Once I swam in the Russian River on the north coast, floating
And silver juvenile steelhead swam between my legs in a deep cool
    pool,
And I watched the sun turn into a night.
Later, we drove down to the redwoods to pitch a tent and light a
    fire
Under the redwoods, late at night, in the silence under the canopy, I
    heard cries that woke me

River is truth. Empty. I am empty. I exist with this river. Nothing.
  Existed.

Near midnight, the sound of crying women filled the air, a circle of
  shared despair, late at night echoing under the canopy of the
  redwoods.

Crying together, like a chant.

My friend and I both awoke—I thought of children who had lost
  their father and who had died while camping under the moon

Women, crying in the darkness, and under the redwoods. Endless
  crying. A circle of crying women.

I started to cry not because of their sadness but their truthfulness,
  and I sobbed with intensity that night.

Sorrow and joy deep, screaming of sadness, in the bone-crushing
  mind-crushing loss I felt with them.

The chorus of weeping women and me.

How could the Buddha smile with this same sorrow I feel, I felt,
  many times over?

Buddha smiles and laughs. I am laughing and crying. I laughed
  while crying.

. . . and the bird calls out under a setting sun.

The last bird. I heard frogs and then silence after spring too. The
  crickets in their darkness. I heard it. I chased that silence at
  night.

Let's call it a bird, said Coyote.

Every morning I hear the first bird's song as if it woke up suddenly.

The bird's cry in the early morning awakens me.

It is perfume, gold, and honey, endless.

I made a nest of redwood cones and slept all day until the sun
  awakened me.

Bibliography

Aars, Jon, Magnus Andersen, Agnès Brenière. and Samuel Blanc. 2015. "White-Beaked Dolphins Trapped in the Ice and Eaten by Polar Bears." *Polar Research* 34:26612.

Aberley, Doug. 1993. *Boundaries of Home: Mapping for Local Empowerment.* Gabriola Island: New Society.

Adger, W. Neil, Jon Barnett, F. S. Chapin III, and Heidi Ellemor. 2011. "This Must Be the Place: Underrepresentation of Identity and Meaning in Climate Change Decision-Making." *Global Environmental Politics* 11(2):1–25.

Adorno, Theodore W. 1984. *Aesthetic Theory.* Trans. C. Lenhardt. London: Routledge & Kegan Paul.

Ainley, David G. 2010. "A History of the Exploitation of the Ross Sea, Antarctica." *Polar Record* 46:233–43.

Allibone, R., B. David, R. Hitchmough, D. Jellyman, N. Ling, P. Ravenscroft, and J. Waters. 2009. "Conservation Status of New Zealand Freshwater Fish." *New Zealand Journal of Marine and Freshwater Research* 44:271–87.

Allison, Graham W., Steven D. Gaines, Jane Lubchenco, and Hugh P. Possingham. 2003. "Ensuring Persistence of Marine Reserves: Catastrophes Require Adopting an Insurance Factor." *Ecological Applications* 13, 1 Supp.: S8–S24.

Anderson, Charles. 2012. "New Zealand's Green Tourism Push Clashes with Realities." *New York Times,* November 16. www.nytimes.com/2012/11/17/business/global/new-zealands-green-tourism-push-clashes-with-realities.html.

Banta, Wendy, and Mark Gibbs. 2009. "Factors Controlling the Development of the Aquaculture Industry in New Zealand: Legislative Reform and Social Carrying Capacity." *Coastal Management* 37:170–96.

Barnes, Jessica, and Michael R. Dove. 2015. *Climate Cultures: Anthropological Perspectives on Climate Change.* New Haven, CT: Yale University Press.

Barnett, Jon. 2003. "Security and Climate Change." *Global Environmental Change* 13:7–17.

Barnett, Jon, and W. Neil Adger. 2007. "Climate Change, Human Security and Violent Conflict." *Political Geography* 26:639–55.

Barnosky, Anthony D. 2008. "Climate Change, Refugia, and Biodiversity: Where Do We Go from Here? An Editorial Comment." *Climate Change* 86:29–32.

———, ed. 2009. *Heatstroke: Nature in an Age of Global Warming.* Covello, CA: Island.

Bartlett, Andrew. 2002. "Refugees & Climate Change: CHOGM Must Prepare to Meet a Changing International Climate" (press release). Australian Democrats. Queensland.

Bartram, Graham, and Anthony Waine, eds. 1982. *Brecht in Perspective.* London: Longman.

Bateson, Gregory. 1972. *Steps to an Ecology of the Mind: Collected Essays in Anthropology, Psychiatry, Evolution, and Epistemology.* San Francisco, CA: Chandler.

Baudrillard, Jean. 1994. *Simulacra and Simulation.* Trans. S. F. Glaser. Ann Arbor: University of Michigan Press, 1994.

Baxter, James K. 1995. "Poem in the Matukituki Valley." In his *Collected Poems,* edited by John Weir. Oxford: Oxford University Press.

Beamish, Thomas D. 2002. *Silent Spill: The Organization of an Industrial Crisis.* Cambridge, MA: MIT Press.

Beaugrand, Grégory, Martin Edwards, Virginie Raybaud, Eric Goberville, and Richard R. Kirby. 2015. "Future Vulnerability of Marine Biodiversity Compared with Contemporary and Past Changes." *Nature Climate Change* 5:695–70. doi:10.1038/nclimate2650

Benjamin, Walter. 1977 [1935]. *The Work of Art in the Age of Mechanical Reproduction: Illuminations.* Trans. H. Zohn. London: Fontana.

Berg, Peter, and Raymond F. Dasmann. 1978. *Reinhabiting a Separate Country: A Bioregional Anthology of Northern California.* San Francisco, CA: Planet Drum Foundation.

Berkes, Fikret. 2012. *Sacred Ecology.* 3rd ed. London: Routledge.

Berry, Thomas. 1982. "Bioregions: The Context for Reinhabiting the Earth." In *Teilhard Studies,* vol. 14. Chambersburg, PA: Anima.

———. 1988. *The Dream of the Earth.* San Francisco, CA: Sierra Club.

Berry, Wendell. 2015. "To Save the Future, Live in the Present." *Yes Magazine,* May 5. www.dailygood.org/story/1037/wendell-berry-on-climate-change-to-save-the-future-live-in-the-present-wendell-berry/.

Blondel, Jacques, and James Aronson. 1999. *Biology and Wildlife of the Mediterranean Region.* London: Oxford University Press.

Blunden, Jessica, and Derek S. Arndt, eds. 2015. "2015: State of the Climate in 2014." *Bulletin American Meteorological Society* 96(7):S1– S267.

Boyce, D. G., M. R. Lewis, and B. Worm. 2003. "Global Phytoplankton Decline over the Past Century." *Nature* 466(7306):591–96.

Bradbury, Ray. 1962. *Something Wicked This Way Comes*. New York: Simon and Schuster.

Bradshaw, C.J.A., X. Giam, and N.S. Sodhi. 2010. "Evaluating the Relative Environmental Impact of Countries." *PLOS One* 5:E10440.

Bremer, Scott, and Bruce Glavovic. 2013. "Exploring the Science-Policy Interface for Integrated Coastal Management in New Zealand." *Ocean & Coastal Management* 84:107–18.

Bronen, Robin. 2012. "Choice and Necessity: Relocations in the Arctic and South Pacific." *Forced Migration Review* 45:17.

Brown, Oli. 2007/2008. *Fighting Climate Change: Human Solidarity in a Divided World*. Occasional paper. Human Development Report, UNDP, Geneva.

Browner, Carol M. 1996. *Watershed Approach Framework*. US Environmental Protection Agency. www.epa.gov/OWOW/watershed/framework.html.

Brownlee, Gerald. 2010. "Funding Boost for New Zealand Aquaculture." *Beehive News Release*, January 19. Wellington, New Zealand.

Buchmann, Stephen L., and Gary P. Nabhan. 1996. *The Forgotten Pollinators*. Washington, DC: Island Shearwater.

Calabresi, Guido, and Philip Bobbitt. 1978. *Tragic Choices*. New York: W.W. Norton.

California Air Resources Board. 2011. "Ocean-Going Vessels Fuel Regulation 1085 Documents." Sacramento, CA. www.arb.ca.gov/ports/marinevess/ogv /ogv1085.htm.

California Coastal Conservancy. 2002. *Regional Wetland Plan*. Oakland, CA: California Resource Agency.

California Department of Fish and Game. 2005. *Final Market Squid Fishery Management Plan*. Sacramento, CA: State of California Resources Agency.

———. 2008a. *California Marine Life Protection Act: Master Plan for Marine Protected Areas*. Approved by the California Fish and Game Commission February 2008.

———. 2008b. "Estimated Long-Term Costs to Implement the California MLPA." In *Master Plan Appendices*, appendix L. Sacramento, California.

———. 2008c. *Channel Islands Marine Protected Areas: First 5 Years of Monitoring: 2003–2008*. Santa Barbara, CA: Partnership for Interdisciplinary Studies of Coastal Oceans, Channel Islands National Marine Sanctuary and Channel Islands National Park.

———. 2011. "California Market Squid." In *California's Living Marine Resources: A Status Report*. Sacramento, CA.

Campbell, J.R. 2010. "Climate-Induced Community Relocation in the Pacific: The Meaning and Importance of Land." In *Climate Change and Migration: Multidisciplinary Perspectives*, ed. J. McAdams, 29–50. Oxford: Hart.

Carr, M.H., G. Forrester, J. Harding, M.V. McGinnis, and P. Raimondi. 2004. *Consequences of Alternative Decommissioning Options to Reef Fish Assemblages and Implications for Decommissioning Policy*. Mineral Management Service, Pacific Region, U.S. Department of the Interior. MMS Cooperative Agreement No. 14–35–0001–30758.

Cato, Molly S. 2012. *The Bioregional Economy: Land, Liberty and the Pursuit of Happiness*. London: Routledge.

Clark, C. W., W. T. Ellison, B. L. Southall, L. Hatch, S. M. Van Parijs, A. Frankel, and D. Ponirakis. 2009. "Acoustic Masking in Marine Ecosystems: Intuitions, Analysis, and Implication." *Marine Ecology Progress Series* 395:201–22.

Code, Lorraine. 2006. *Ecological Thinking: The Politics of Epistemic Location.* Oxford: Oxford University Press.

Cohen, Andrew, and James Carlton. 1997. "Transoceanic Transport Mechanisms: Introduction of the Chinese Mitten Crab, *Eriocheir sinensis*, to California." *Pacific Science* 51:1–11.

Costanza, R., L. Graumich, and W. Steffens, eds. 2006. *Integrated History and Future of People on Earth.* Cambridge, MA: MIT Press.

Dailey, Murray D., Donald J. Reish, and Jack W. Anderson, eds. 1993. *Ecology of the Southern California Bight: A Synthesis and Interpretation.* Berkeley: University of California Press.

DairyNZ. 2011/2012. *New Zealand Dairy Statistics.* Wellington, NZ. www .dairynz.co.nz/media/434165/new_zealand_dairy_statistics_2011–12.pdf

Dark Mountain Project. 2015. *Uncivilisation: The Dark Mountain Manifesto.* http://dark-mountain.net/about/manifesto/.

Darnell, Rezneat M., and Richard E. Defenbaugh. 1990. "Gulf of Mexico: Environmental Overview and History of Environmental Research." *American Zoologist* 30(1):3–6.

Dasmann, Raymond. 1975 (February). "National Parks, Nature Conservation, and Future Primitive." Presented at the South Pacific Conference on National Parks, Wellington, New Zealand.

Daston, Laura. 1998. "Fear and Loathing of the Imagination in Science." *Daedalus* 127:73–95.

Davenport, Coral. 2014. "Rising Seas." *New York Times,* March 27. www .nytimes.com/interactive/2014/03/27/world/climate-rising-seas.html.

Davis, Gary E. 2005. "Science and Society: Marine Reserve Design for the California Channel Islands." *Conservation Biology* 19:1745–51.

Davis, Mike. 1991. *City of Quartz: Excavating the Future of Los Angeles.* New York: Verso.

———. 1995. "How Eden Lost Its Garden: A Political History of the L.A. Landscape." *Capitalism Nature Socialism* 6(4):1–29.

Davis, Wade. 2007. "Dreams from Endangered Cultures." TED Talk video. www.ted.com/talks/wade_davis_on_endangered_cultures.

Dickens, Charles. 1850. "A Preliminary Word." In *Household Words.* www .victorianweb.org/periodicals/hw.html.

Diamond, Jared. 2005. *Collapse: How Societies Choose to Fail or Succeed.* New York: Viking.

Douvere, Fanny. 2008. "The Importance of Marine Spatial Planning in Advancing Ecosystem-Based Sea Use Management." *Marine Policy* 32:762–71.

Dunlap, R. E., K. D. Van Liere, A. G. Mertig, and R. E. Jones. 2000. "Measuring Endorsement of the New Ecological Paradigm: A Revised NEP Scale." *Journal of Social Issues* 56(3):425–42.

Eddy, Tyler E. 2014. "One-Hundred-Fold Difference between Perceived and Actual Levels of Marine Protection in New Zealand." *Marine Policy* 46: 61–67.

Egan, David. 2008. "Climate Change Fuels Forced Migration." *Measurement & Human Rights Issue Paper* 1, no. 2: 1–2.

Ehler, Charles, and Fanny Douvere. 2007. *Visions for a Sea Change: Report of the First International Workshop on Marine Spatial Planning.* Intergovernmental Oceanographic Commission and Man and the Biosphere Programme, IOC Manual and Guides, 46: ICAM Dossier, 3. Paris: UNESCO.

———. 2009. *Marine Spatial Planning: A Step-by-Step Approach toward Ecosystem-Based Management.* Intergovernmental Oceanographic Commission and Man and the Biosphere Programme. Paris: UNESCO.

Elder, Jon, ed. 1998. *Stories in the Land: A Place-Based Environmental Education Anthology.* Great Barrington, MA: Orion Society.

Ericksen, Neil F. 1990. "New Zealand Water Planning and Management: Evolution or Revolution?" In *Integrated Water Management: International Experiences and Perspectives,* ed. B. Mitchell. London: Belhaven.

Erlandson, J.M., T.C. Rick, T.J. Braje, M. Casperson, B. Culleton, B. Fulfrost, T. Garcia, D.A. Guthrie, N. Jew, D.J. Kennett, M.L. Moss, L. Reeder, C. Skinner, J. Watts, and L. Willis. 2011. "Paleoindian Seafaring, Maritime Technologies, and Coastal Foraging on California's Channel Islands." *Science* 331:1181–85.

Evanoff, Richard. 2011. *Bioregionalism and Global Ethics.* London, UK: Routledge.

Fagan, Brian. 2004. *The Long Summer: How Climate Changed Civilization.* New York: Perseus.

Farady, Susan E. 2006. "Compatible Use within National Marine Sanctuaries: Determining Meaningful Implementation." *Ocean and Coastal Law Journal* 12:1–13.

Feely, R., C.L. Sabine, J.M. Hernandez-Ayon, D. Ianson, and B. Hales. 2008. "Evidence for Upwelling of Corrosive 'Acidified' Seawater onto the Continental Shelf." *Science* 320(5882):1490–92.

Finlayson, C.M., Rebecca D'Cruz, and Nick Davidson. 2005. *Ecosystems and Human Well-Being: Wetlands and Water: Synthesis.* Washington, DC: World Resources Institute.

Firth, Raymond. 1940. "The Analysis of Mana: An Empirical Approach." *Journal of the Polynesian Society* 49(196):482–510.

Foley, M.M., B.S. Halpern, F. Micheli, M.H. Armsby, M.R. Caldwell, C.M. Crain, E. Prahler, N. Rohr, D. Sivas, M.W. Beck, M.H. Carr, L.B. Crowder, J.E. Duffy, S.D. Hacker, K.L. McLeod, S.R. Palumbi, C.H. Peterson, H.M. Regan, M.H. Ruckelshaus, P.A. Sandifer, and R.S. Steneck. 2010. "Guiding Ecological Principles for Marine Spatial Planning." *Marine Policy* 34:955–66.

Foote, K.J., M.K. Joy, and R.G. Death. 2015. "New Zealand Dairy Farming: Milking Our Environment for All Its Worth." *Environmental Management* 56(3):709–20.

Forrest, B., N. Keeley, P. Gillespie, G. Hopkins, B. Knight, and D. Govier. 2007. *Review of the Ecological Effects of Marine Finfish Aquaculture: Final Report.* Cawthron Report no. 1285. Nelson, New Zealand: Ministry of Fisheries.

Freudenberg, William R., and Robert Gramling. 1994. *Oil in Troubled Waters: Perceptions, Politics, and the Battle over Offshore Drilling.* New York: State University of New York Press.

George, Rose. 2013. *Ninety Percent of Everything: Inside Shipping, the Invisible Industry that Puts Clothes on Your Back, Gas in Your, and Food on Your Plate.* New York: Metropolitan.

Gibbs, Mark. 2010. "The Historical Development of Fisheries in New Zealand with Respect to Sustainable Development Principles." *Electronic Journal of Sustainable Development* 1(2):23–33.

Gibbs, M. T., J. Hobday, B. Sanderson, and C. L. Hewitt. 2006. "Defining the Seaward Extent of New Zealand's Coastal Zone." *Estuarine, Coastal and Shelf Science* 66:240.

Gleason, M., S. McCreary, M. Miller-Henson, J. Ugoretz, E. Fox, M. Merrifield, W. McClintock, P. Serpa, and K. Hoffman. 2010. "Science-Based and Stakeholder-Driven Marine Protected Area Network Planning: A Successful Case Study from North Central California." *Ocean & Coastal Management* 53(2):52–68.

Glotflety, Cheryll, and Eve Quesnel, eds. 2015. *The Biosphere and the Bioregion: Essential Writings of Peter Berg.* London: Routledge Environmental Humanities.

Gopnik, Morgan. 2008. *Integrated Marine Spatial Planning in US Waters: The Path Forward.* Marine Conservation Initiative. Palo Alto, CA: Gordon and Betty Moore Foundation.

Gordon, Dennis P., Jennifer Beaumont, Alison MacDiarmid, Donald A. Robertson, and Shane T. Ahyong. 2010. "Marine Biodiversity of Aotearoa New Zealand." *PLoS ONE* 5(8): e10905. doi:10.1371/journal.pone.0010905

Graynoth, Eric. 2006. "The Long and the Short of It: Looking After the Needs of Native Eels." *Details* 14(12):22–23. National Institute of Water and Atmospheric Research.

Greenberg, Paul. 2014. *American Catch: The Fight for Our Local Seafood.* New York: Penguin.

Griffin, Daniel, and Kevin J. Anchukaitis. 2014. "How Unusual is the 2012–2014 California Drought?" *Geophysical Research Letters* 41: 9017–23.

Grim, John A., ed. 2001. *Indigenous Traditions and Ecology: The Interbeing of Cosmology and Community.* Cambridge, MA: Harvard University Press.

Halpern, Benjamin S. 2014. "Conservation: Making Marine Protected Areas Work." *Nature* 506:167–68.

Halpern, Benjamin S., Sarah E. Lester, and Karen L. Mcleod. 2010. "Marine Reserves Special Feature: Placing Marine Protected Areas onto the Ecosystem-based Management Seascape." *Proceedings of the National Academy of Sciences* 107:18312–17.

Halpern, Benjamin S., Karen McLeod, Andrew Rosenberg, and Larry Crowder. 2008. "Understanding Cumulative and Interactive Impacts as a Basis for Ecosystem-Based Management and Ocean Zoning." *Ocean and Coastal Management* 51:203–11.

Hansen, J., L. Nazarenko, R. Ruedy, M. Sato, J. Willis, A. Del Genio, D. Koch, A. Lacis, K. Lo, S. Menon, T. Novakov, J. Perlwitz, G. Russell, G. A. Schmidt, and N. Tausnev. 2005. "Earth's Energy Imbalance: Confirmation and Implications." *Science* 308:1431–35.

Haraway, Donna. 2008. "Situated Knowledges: The Science Question in Feminism and the Privilege of Partial Perspective." *Feminist Studies* 14(3):575–99.

Harding, Garrett. 1974 (September). "Lifeboat Ethics: The Case Against Helping the Poor." *Psychology Today* 8:148–60.

Harding, Russell. 2007. "Muddying the Waters: Managing Water Quality in New Zealand." *Policy Quarterly* 3(3):16–24.

Harms, S., and C. D. Winant. 1998. "Characteristic Patterns of the Circulation in the Santa Barbara Channel." *Journal of Geophysical Research* 103(C2):3041–65.

Harrison, Robert P. 1996. "Toward a Philosophy of Nature." In *Uncommon Ground: Rethinking the Human Place in Nature*, edited by William Cronon, 426–37. New York: W. W. Norton.

Harville, J. P. 1983. "Obsolete Petroleum Platforms as Artificial Reefs." *Fisheries* 8(2):4–6.

Hayman, Ronald. 1977. *Artaud and After*. Oxford: Oxford University Press.

Hazen, T. C. 2011. *FAQ: Microbes & Spills. A Report from the American Academy of Microbiology*. Washington, DC. http://academy.asm.org/images /stories/documents/Microbes_and_Oil_Spills.pdf.

Heaney, Seamus. 1966. "Digging." In *Death of a Naturalist*. New York: Farrar, Straus & Giroux.

Hecht, Sean B. 2010. *California's New Rigs-to-Reefs Law*. UCLA Institute of the Environment and Sustainability, School of Law. www.environment.ucla .edu/reportcard/article9389.html.

Helvey, Mark. 2004. "Seeking Consensus on Designing Marine Protected Areas: Keeping the Fishing Community Engaged." *Coastal Management* 32:173–90.

Hilborn, Richard. 2011 (February). "Five Myths." Presentation at Victoria University of Wellington, New Zealand.

Hoekstra, Arjen Y. 2012. "The Hidden Water Resource Use behind Meat and Dairy." *Animal Frontiers* 2(2):1–8.

Holbrook, Sally J., Richard F. Ambrose, Louis Botsford, Mark H. Carr, Peter T. Raimondi, and Mia J. Tegner. 2000. *Ecological Issues Related to Decommissioning of California's Offshore Production Platforms*. Report to the University of California Marine Council, Berkeley.

Holtved, Erik. 1966/1967. "The Eskimo Myth about the Sea-Woman: A Folkloristic Sketch." *Folk* 8/9:145–53.

Hong, Sun-Kee. 2011. "Eco-cultural Diversity in Island and Coastal Landscapes: Conservation and Development." In *Landscape Ecology in Asian Cultures*, ed. S. K. Hong, J. Wu, J. E. Kim, and N. Nakagoshi, 11–28. Tokyo: Springer.

———. 2013. "Biocultural Diversity Conservation for Island and Islanders: Necessity, Goal and Activity." *Journal of Marine and Island Cultures* 2:102–06.

Hong, S., Maffi, L., Oviedo, G., Matsuda, H., and Kim, J. E. 2013. "Island Biocultural Diversity Initiative." *INTECOL E-bulletin* 7 (March), 7–9.

Horwitz, Joshua. 2014. *War of the Whales*. New York: Simon & Schuster.

House, Freeman. 1999. *Totem Salmon: Lessons from Another Species*. Boston: Beacon.

Irvine, L. M., B. R. Mate, M. H. Winsor, D. M. Palacios, S. J. Bograd, D. P. Costa, and H. Bailey. 2014. "Spatial and Temporal Occurrence of Blue Whales off the U.S. West Coast, with Implications for Management." *PLoS ONE* 9(7):e102959.

Jackson, J. B. C., M. X. Kirby, W. H. Berger, K. A. Bjorndal, L. W. Botsford, B. J. Bourque, R. H. Bradbury, R. Cooke, J. Erlandson, J. A. Estes, T. P. Hughes, S. Kidwell, C. B. Lange, H. S. Lenihan, J. M. Pandolfi, C. H. Peterson, R. S. Steneck, M. J. Tegner, and R. R. Warner. 2001. "Historical Overfishing and the Recent Collapse of Coastal Ecosystems." *Science* 293(5530):629–37.

Jackson, Jeremy B. C. 2001. "What Was Natural in the Coastal Oceans?" *Proceedings of the National Academy of Sciences* 98: 5411–18. doi:10.1073/pnas.091092898

Johnson, John R. 2000. "Social Responses to Climate Change among Chumash Indians of South Central California." In *The Way the Wind Blows: Climate, History, and Human Action*, ed. R. J. McIntosh, J. A. Tainter, and S. K. McIntosh, 301–27. New York: Columbia University Press.

Joy, Mike. 2011. "The Dying Myth of a Clean, Green Aotearoa." *New Zealand Herald*, April 25. www.nzherald.co.nz/business/news/article.cfm?c_id = 3&objectid = 10721337.

Kahui, Viktoria, and Amanda C. Richards. 2014. "Lessons from Resource Management by Indigenous Maori in New Zealand: Governing the Ecosystems as a Commons." *Ecological Economics* 102:1–7.

Kawharu, Merata. 1998. *Dimensions of Kaitiakitanga: An Investigation of a Customary Māori Principle of Resource Management*. PhD thesis, Oxford University.

Kennett, James P., and B. Lynn Ingram. 1995. "A 20,000-Year Record of Ocean Circulation and Climate Change from the Santa Barbara Basin." *Nature* 377:510–14.

Kibel, Paul S. 2014. "A Salmon Eye Lens on Climate Adaption." *Ocean and Coastal Law Journal* 19:65–91.

King, Chester. 1990. *Evolution of Chumash Society: A Comparative Study of Artifacts Used in Social System Maintenance in the Santa Barbara Channel Region Before AD 1804*. New York: Garland.

Kingsford, Richard T., and James E. M. Watson. 2011. "Climate Change in Oceania: A synthesis of Biodiversity Impacts and Adaptations." *Pacific Conservation Biology* 17:270–84.

Kingsford, R. T., J. E. M. Watson, C. J. Lundquist, O. Venter, L. Hughes, E. L. Johnston, J. Atherton, M. Gawel, D. A. Keith, B. G. Mackey, C. Morley, H. P. Possingham, B. Raynor, H. F. Recher, and K. A. Wilson. 2009. "Major Conservation Policy Issues for Biodiversity in Oceania." *Conservation Biology* 23:834–40.

Klausmeyer, Kirk R., and M. Rebecca Shaw. 2009. "Climate Change, Habitat Loss, Protected Areas and the Climate Adaptation Potential of Species in Mediterranean Ecosystems Worldwide." *PLoS ONE* 4(7):e6392. doi:10.1371/journal.pone.0006392

Knodel, Marissa S. 2012. "Wet Feet Marching: Climate Justice and Sustainable Development for Climate Displaced Nations in the South Pacific." *Vermont Journal of Environmental Law* 14:127–76.

Kummu, Matti, Hans de Moel, Philip J. Ward, and Olli Varis. 2011. "How Close Do We Live to Water? A Global Analysis of Population Distance to Freshwater Bodies." *PLoS ONE* 6(6): e20578. doi:10.1371/journal.pone.0020578

Lazarus, Richard J. 2009. "Super Wicked Problems and Climate Change: Restraining the Present to Liberate the Future." *Cornell Law Review* 94:1153–1234.

Le Clézio, Jean-Marie-Gustave. 2007. *Raga: Approche du continent invisible.* Paris: Editions Du Seuil.

Leaper, Russel, Martin Renilson, and Conor Ryan. 2014. "Reducing Underwater Noise from Large Commercial Ships: Current Status and Future Directions." *Journal of Ocean Technology* 9:51.

Leathwick, John, Atte Moilanen, Malcolm Francis, Jane Elith, Paul Taylor, Kathryn Julian, Trevor Hastie, and Clinton Duffy. 2008. "Novel Methods for the Design and Evaluation of Marine Protected Areas in Offshore Waters." *Conservation Letters* 1:96–99.

Leet, William S. 2001. *California's Living Marine Resources: A Status Report.* Sacramento: California Department of Fish and Game.

Leopold, Aldo. 2001. *A Sand County Almanac.* Oxford: Oxford University Press.

Lester, S. E., K. L. Mcleod, H. Tallis, M. Ruckelshaus, B. S. Halpern, P. S. Levin, F. P. Chavez, C. Pomeroy, B. J. Mccay, C. Costello, S. D. Gaines, A. J. Mace, J. A. Barth, D. L. Fluharty, and J. K. Parrish. 2010. "Science in Support of Ecosystem-Based Management for the US West Coast and Beyond." *Biological Conservation* 143(3):576–87.

Lévi-Strauss, Claude. 1978. *Myth and Meaning.* Toronto: University of Toronto Press.

Levin, K., B. Cashone, S. Bernstein, and G. Auld. 2012. "Overcoming the Tragedy of Super Wicked Problems: Constraining Our Future Selves to Ameliorate Global Climate Change." *Policy Science* 45:123–52.

Lieber, Michael D. 1978. *Exiles and Migrants in Oceania.* Association for Social Anthropology in Oceania Monograph 5. Honolulu: University of Hawaii Press.

Linklater, Wayne. 2012. "Maui's dolphins need better protection—scientists." Posted in Science Alert: Experts Respond. www.sciencemediacentre.co.nz/2012/03/14/mauis-dolphins-need-better-protection-scientists/.

Loarie, Scott R., Benjamin E. Carter, Katharine Hayhoe, Sean McMahon, Richard Moe, Charles A. Knight, and David D. Ackerly. 2008. "Climate Change and the Future of California's Endemic Flora." *PLoS ONE* 3(6):e2502. doi:10.1371/journal.pone.0002502

Lopez, Barry. 1990. *Crow and Weasel.* San Francisco, CA: North Point Press.

Louv, Richard. 2005. *Last Child in the Woods: Saving Our Children from Nature-Deficit Disorder.* Chapel Hill, NC: Algonquin of Chapel Hill.

Love, Milton S., Jennifer E. Caselle, and William Van Buskirk. 1998. "A Severe Decline in the Commercial Passengers Fishing Vessel Rockfish (*Sebastes sp.*)

Catch in the Southern California Bight, 1980–1996." CalCOFI Report 39. San Diego, CA: California Cooperative Oceanic Fisheries Investigations.

Love, Milton S., Jennifer Caselle, and Larry Snook. 1999. "Fish Assemblages on Mussel Mounds Surrounding Seven Oil Platforms in the Santa Barbara Channel and Santa Maria Basin." *Bulletin of Marine Science* 65(2):497–513.

Lubchenco, Jane, and Laura Petes. 2010. "The Interconnected Biosphere: Science at the Ocean's Tipping Points." *Oceanography* 23(2)115–29.

Luccarelli, Mark. 1995. *Lewis Mumford and the Ecological Region: The Politics of Planning.* New York: Guilford Press.

Lynch, Tom, Cheryll Glotfelty, and Karla Armbruster, eds. 2012. *The Bioregional Imagination: Literature, Ecology, and Place.* Athens: University of Georgia Press.

MacDiarmid, A., J. Beaumont, H. Bostock, D. Bowden, M. Clark, M. Hadfield, P. Heath, G. Lamarche, S. Nodder, A. Orpin, C. Stevens, D. Thompson, L. Torres, and R. Wysoczanski. 2012. *Expert Risk Assessment of Activities in the New Zealand Exclusive Economic Zone and Extended Continental Shelf.* NIWA Client Report No. WLG2011–39. Wellington, New Zealand: Ministry for the Environment.

MacDiarmid, A., C. Law, M. Pinkerton, and J. Zeldis. 2013. "New Zealand Marine Ecosystem Services." In *Ecosystem Services in New Zealand: Conditions and Trends,* ed. J. Dymond, 238–53. Lincoln: Manaaki Whenua Press.

MacDiarmid, A., A. McKenzie, J. Sturman, J. Beaumont, S. Mikaloff-Fletcher, and J. Dunne. 2012. *Assessment of Anthropogenic Threats to New Zealand Marine Habitats.* New Zealand Aquatic Environment and Biodiversity Report no. 93. Wellington: Ministry of Agriculture and Forestry.

Maffi, Louisa. 1998. "Language: A Resource for Nature. Nature and Resources." *UNESCO Journal of Environment and Natural Resource Research* 34(4):12–21.

———. 2008. "Talking diversity." *World Conservation* (January):13–14.

Maffi, Louisa, and Ellen Woodley. 2010. *Biocultural Diversity Conservation: A Global Sourcebook.* London: Earthscan.

Makgill, Robert A., and Hamish G. Rennie. 2012. "A Model for Integrated Coastal Management Legislation: A Principled Analysis of New Zealand's Resource Management Act 1991." *International Journal of Marine and Coastal Law* 27:135–65.

Marine Reserves Working Group, Science Advisory Panel. 2001. "How Large Should Marine Reserve Be?" Santa Barbara, CA: Channel Islands National Marine Sanctuary Program.

Marx, Leo. 2000. *The Machine in the Garden: Technology and the Pastoral Idea in America.* Oxford: Oxford University Press.

McCauley, D.J., M.L. Pinksy, S.R. Palumbi, J.A. Estes, F.H. Joyce, and R.R. Warner. 2015. "Marine Defaunation: Animal Loss in the Global Ocean." *Science* 347(6219):247–54.

McEvoy, Arthur F. 1986. *The Fisherman's Problem: Ecology and Law in the California Fisheries, 1850–1980.* Cambridge: Cambridge University Press.

McGinnis, Michael V. 1994. "The Politics of Restocking vs. Restoring Salmon in the Columbia River Basin." *Restoration Ecology* 2(3):149–55.

———. 1995. "On the Verge of Collapse: The Columbia River System, Wild Salmon and the Northwest Power Planning Council." *Natural Resources Journal* 35:63–92.

———. 1998. "An Analysis of the Role of Ecological Science in Offshore Continental Shelf Abandonment Policy." In *Taking a Look at California Ocean Resources: An Agenda for the Future*, ed. O.T. Magoon, H. Converse, B. Baird, and M. Miller-Henson, 1384–92. Reston, VA: American Society of Civil Engineers.

———, ed. 1999a. *Bioregionalism*. London: Routledge.

———. 1999b. "Re-wilding Imagination: Mimesis and Ecological Restoration." *Ecological Restoration* 17(4):219–26.

———. 2006. "Negotiating Ecology: Marine Bioregions and the Destruction of the Southern California Bight." *Futures* 38:400–01.

———. 2011. "Mindfulness of the Oceanic Commons." *Pacific Ecologist* 20:55–60.

———. 2012a. "Learning from California's Experience in Marine Life Protection." *Ocean Yearbook* 26:485–508.

———. 2012b. "Living up to the Brand: Greening New Zealand's Ocean Policy." *Policy Quarterly* 8:17–28.

———. 2012c. *Marine Governance: The New Zealand Dimension. Full Report.* Funded by the Ministries of New Zealand. Emerging Issues Program, School of Government, Victoria University of Wellington, New Zealand.

McGinnis, Michael V., and Megan Collins. 2013. "A Race for Marine Space: Science, Values, and Aquaculture Planning in New Zealand." *Coastal Management* 41(5):401–19.

McGinnis, Michael V., and Roberta R. Cordero. 2004. *Tribal Marine Protected Areas: Protecting Maritime Ways and Cultural Practices.* Santa Barbara, CA: Bioregional Planning Associates.

McGinnis, Michael V., Linda Fernandez, and Carolyn Pomeroy. 2001. *The Politics, Ecology and Economics of Decommissioning of California Offshore Oil and Gas Structures.* Santa Barbara, CA: Mineral Management Service, Pacific Region, U.S. Department of the Interior.

McGinnis, Michael V., Freeman House, and William R. Jordan III. 1999. "Bioregional Restoration: Re-establishing an Ecology of Shared Identity." In *Bioregionalism*, ed. M.V. McGinnis, 205–22. London: Routledge.

McGinnis, Michael V., and John T. Woolley. 1997. "The Discourses in Restoration." *Restoration and Management Notes* 15:74–77.

McGinnis, Michael V., John T. Woolley, and John K. Gamman. 1999. "Bioregional Conflict Resolution: Rebuilding Community in Watershed-based Planning and Organizing." *Environmental Management* 24:1–12.

McGinnis, Michael V., John T. Woolley, and William M. Herms. 1999. "Survey Methodologies for the Study of Ecosystem Restoration and Management: The Importance of Q-Methodology." In *Integrated Assessment of Ecosystem Health*, ed. Kate M. Scow et al., 321–32. Boca Raton, FL: Lewis.

McIntosh, Angus R., and Colin R. Townsend. 1995. "Contrasting Predation Risks Presented by Introduced Brown Trout and Native Common River Galaxias in New Zealand Streams." *Canadian Journal of Fisheries and Aquatic Sciences* 52:1821–33.

McKenna, M.F., S.L. Katz, C. Condit, and S. Walbridge. 2012. "Response of Commercial Ships to a Voluntary Reduction Measure: Are Voluntary Strategies Adequate for Mitigating Ship-Strike Risk?" *Coastal Management* 40(6):634–50.

McKenna, M.F., S.L. Katz, S.M. Wiggins, D. Ross, and J.A. Hildebrand. 2012. "A Quieting Ocean: Unintended Consequence of a Fluctuating Economy." *Journal of the Acoustical Society of America* 132(3):EL169–EL175.

McKenna, M.F., S.M. Wiggins, and J.A. Hildebrand. 2013. "Relationship between Container Ship Underwater Noise Levels and Ship Design, Operational and Oceanographic Conditions." *Scientific Reports* 3:1760.

McNamee, Gregory. 1994. *Gila: The Life and Death of an American River.* New York: Orion.

Meakins, Brook. 2012. "Village Relocated Due to Climate Change." *Salon,* September 19. www.salon.com/2012/09/19/first_village_relocated_due_to_climate_change/.

Mekonnen, Mesfin M., and Arjen Y. Hoekstra. 2011. *National Water Footprint Accounts: The Green, Blue and Gray Water Footprint of Production and Consumption.* Value of Water Resources Report Series no. 50. Delft, the Netherlands: UNESCO-IHE.

Melville, Herman. 1951. *Moby-Dick, or The Whale.* New York: Harper and Brothers.

Merleau-Ponty, Maurice. 1964. *The Primacy of Perception,* ed. and intro. by J.M. Edie. Evanston, IL: Northwestern University Press.

Miles, Edward L. 2009. "On the Increasing Vulnerability of the World Ocean to Multiple Stresses." *Annual Review of Environment and Resources* 34:17–41.

Ministry for the Environment. 2007. *Improving Regulation of Environmental Effects in New Zealand's Exclusive Economic Zone.* Wellington: Ministry for the Environment.

———. 2012. *The Natural Resources Sector: Briefing to Incoming Ministers.* Wellington, New Zealand.

Moffitt, Sarah E., Tessa M. Hill, Peter D. Roopnarine, and James P. Kennett. 2015. "Response of Seafloor Ecosystems to Abrupt Global Climate Change." *Proceedings of the National Academy of Sciences* 112(15):4684–89.

Monnahan, C.C., T.A. Branch, and A.E. Punt. 2014. "Do Ship Strikes Threaten the Recovery of Endangered Eastern North Pacific Blue Whales?" *Marine Mammal Science* 31(1):279–97.

Monnahan, C.C., T.A. Branch, K.M. Stafford, Y.V. Ivashchenko, and E.M. Oleson. 2014. "Estimating Historical Eastern North Pacific Blue Whale Catches Using Spatial Calling Patterns." *PLoS ONE* 9(6):e98974. doi:10.1371/journal.pone.0098974

Morrison, A., M.L. Lowe, D.M. Parsons; N.R. Usmar, and I.M. Mcleod. 2009. "A Review of Land-Based Effects on Coastal Fisheries and Supporting

Biodiversity in New Zealand." New Zealand Aquatic Environment and Biodiversity Report no. 37. Wellington, NZ.

Mumford, Lewis. 1956. "The Natural History of Urbanization." In *Man's Role in the Changing Face of the Earth,* ed. W.L. Thomas, 382–98. Chicago, IL: University of Chicago Press.

———. 1961. *The City in History: Its Origins, Its Transformations, and its Prospects.* New York: Harcourt.

Murray, Catherine, and Garry McDonald. 2010. *Aquaculture: Economic Impact in the Auckland Region.* Auckland Regional Council Document. Technical Report no. 9. Jointly prepared by the Auckland Regional Council and Market Economics Ltd. for Auckland Regional Council.

Murray, J.D. 1994. "A Policy and Management Assessment of U.S. Artificial Reef Programs." *Bulletin of Marine Science* 55(2–3):960–69.

Naeem, Shahid, Daniel E. Bunker, Andy Hector, Michel Loreau, and Charles Perrings, eds. 2009. *Biodiversity, Ecosystem Functioning, and Human Wellbeing: An Ecological and Economic Perspective.* Oxford: Oxford University Press.

National Research Council. 1996. *An Assessment of Techniques for Removing Offshore Structures.* 1996. Washington, DC: National Academies Press.

Nixon, Richard. 2011a. "Slow Violence." *Chronicle Review,* June 26. http://chronicle.com/article/Slow-Violence/127968/.

———. 2011b. *Slow Violence and the Environmentalism of the Poor.* Cambridge, MA: Harvard University Press.

Norse, Elliott A. 2010. "Ecosystem-Based Spatial Planning and Management of Marine Fisheries: Why and How?" *Bulletin of Marine Science* 86(2):179–95.

Oceans Policy Secretariat. 2003a. *Setting the Scene: New Zealand's Oceans-Related Obligations and Work on the International Stage.* Working Paper 1. Wellington: Ministry for the Environment.

———. 2003c. *International Ocean Issues.* Working paper 11. Wellington: Ministry for the Environment.

O'Riordan, Timothy. 1995. "Frameworks for Choice: Core Beliefs and the Environment." *Environment* 37(8):4–29.

Orr, David W. 1992. *Ecological Literacy: Education and the Transition to a Postmodern World.* Albany: State University of New York.

Osmond, Michael, Satie Airame, Meg Caldwell, and Jon Day. 2010. "Lessons for Marine Conservation Planning: A Comparison of Three Marine Protected Area Planning Processes." *Ocean & Coastal Management* 53(2): 41–51.

Ott, Riki. 2008. *Not One Drop: Betrayal and Courage in the Wake of the Exxon Valdez Oil Spill.* White River Junction, VT: Chelsea Green.

Park, Geoff. 1995. *Nga Uruora: Ecology and History in a New Zealand Landscape.* Wellington: Victoria University Press.

———. 2006. *Theatre Country: Essays on Landscape and Whenua.* Wellington: Victoria University Press.

Peart, Raewyn. 2011. *Governing Our Oceans: Environmental Reform for the Exclusive Economic Zone.* Auckland: Environmental Defence Society.

Peninsula Principles on Climate Displacement within States. 2013. Victoria, Australia. http://displacementsolutions.org/wp-content/uploads/FINAL-Peninsula-Principles-FINAL.pdf.

Pietz, David. 2002. *Engineering the State: The Huai River and Reconstruction in Nationalist China, 1927–37*. London: Routledge.

Popkin, Barry M., and Shufa Du. 2003. "Dynamics of the Nutrition Transition toward the Animal Foods Sector in China and Its Implications: A Worried Perspective." *Journal of Nutrition* 13(3):898S–906S.

Post, E., U.S. Bhatt, C.M. Bitz, J.F. Brodie, T.L. Fulton, M. Hebblewhite, J. Kerby, S.J. Kutz, I. Stirling, and D.A. Walker. 2013. "Ecological Consequences of Sea-Ice Decline." *Science* 341:519–24.

Priest, D.F. 1993. *California's 1987–92 Drought: A Summary of Six Years of Drought*. Sacramento: Department of Water Resources State of California.

Prosek, James. 2010. "Maori Eels: New Zealand's Maori Defend an Extraordinary Creature—and Themselves." *Orion*, July/August. https://orionmagazine.org/article/survivors/.

Raab, L. Mark, and Terry Jones. 2004. *Prehistoric California*. Salt Lake City: University of Utah Press.

Reggio, V.C., Jr. 1987a. "Rigs-to-Reefs." *Fisheries* 12(4):2–7.

———. 1987b. *Rigs-to-Reefs: The Use of Obsolete Petroleum Structures as Artificial Reefs*. OCS Report/MMS 87–0015. Minerals Management Service, Gulf of Mexico OCS Regional Office, Metairie, Louisiana.

Reggio, V.C., Jr., and R. Kasprzak. 1991. "Rigs-to-Reefs: Fuel for Fisheries Enhancement through Cooperation." *American Fisheries Society Symposium* 11:9–17.

Roberts, M., W. Norman, N. Minhinnick, D. Wihongi, and C. Kirkwood. 1995. "Kaitiakitanga: Maori Perspectives on Conservation." *Pacific Conservation Biology* 2:7–20.

Robertson, David. 1996. "Bioregionalism in Nature Writing." In *American Nature Writers*, vol. 2. New York: Charles Scribner's Sons.

Rosenberg, A.A., and P.A. Sandifer. 2009. "What Do Managers Need?" In *Ecosystem-Based Management for the Oceans*, ed. Karen McLeod and Heather Leslie, 13–30. Washington, DC: Island Press.

Rothenberg, David. 1996. "No World but in Things: The Poetry of Naess's Concrete Contents." *Inquiry* 39(2):255–72.

Rothwell, Donald R., and Tim Stephens. 2010. *The International Law of the Sea*. Portland, OR: Hart.

Royal Society of New Zealand. 2011 (July). *Ecosystem Services: Emerging Issues*. Ecosystem Services in Policy Workshop, 1–6.

———. 2012 (May). *Future Marine Resource Use: Emerging Issues*.

Rundel, Philip W., Gloria Montenegro, and Fabian Jaksic, eds. 1998. *Landscape Disturbance and Biodiversity in Mediterranean-type Ecosystems*. Berlin: Springer.

Ruru, Jacinta. 2009. *The Legal Voice of Māori in Freshwater Governance: A Literature Review*. New Zealand: Landcare Research.

Saarman, Emily T., and Mark H. Carr. 2013. "The California Marine Life Protection Act: A Balance of Top Down and Bottom Up Governance in MPA Planning." *Marine Policy* 41:41–49.

Safina, Carl. 2010 (May 13). "No Island Is an Island." http://carlsafina.org/no-island-is-an-island/.

Sale, Kirkpatrick. 1985. *Dwellers in the Land: The Bioregional Vision*. San Francisco, CA: Sierra Club.

Sax, Dov F., and Steven D. Gaines. 2008. "Species Invasions and Extinction: The Future of Native Biodiversity on Islands." *Proceedings of the National Academy of Sciences* 105, supp. 1: 11490–97.

Schattsneider, E. E. 1960. *The Semi-Sovereign People : A Realist's View of Democracy in America*. New York: Viking.

Schlosberg, David. 2012. "Climate Justice and Capabilities: A Framework for Adaptive Policy." *Ethics and International Affairs* 26(4):445–61.

Schmitt, Carl. 1950. *The Nomos of the Earth in the International Law of The Jus Publicum Europaeum*. Trans. G. L. Ulmen. New York: Telos.

Schroeder, Donna A., and Milton S. Love. 2004. "Ecological and Political Issues Surrounding Decommissioning of Offshore Oil Facilities in the Southern California Bight." *Ocean and Coastal Management* 47:21–48.

Schubert, R., H. Schellnhuber, N. Buchmann, A. Epiney, R. Grießhammer, M. Kulessa, D. Messner, S. Rahmstorf, and J. Schmid. 2014. *The Future Oceans: Warming up, Rising High, Turning Sour*. Special Report. Berlin: German Advisory Council on Global Change.

Secretariat of the Pacific Regional Environment Programme. 2014. "Funding." www.sprep.org/Funding/funding.

Selby, R., P. Moore, and M. Mulholland. 2010. *Māori and the Environment: Kaitiaki*. Wellington, NZ: Huia Books.

Shafer, C., G. Inglis, and V. Martin. 2010. "Examining Residents' Proximity, Recreational Use, and Perceptions Regarding Proposed Aquaculture Development." *Coastal Management* 38(5):559–74.

Shepard, Paul. 1996. *The Others: How Animals Made Us Human*. Washington, DC: Island.

Shrader-Frechette, Kristin S., and Earl D. McCoy. 1994. *Method in Ecology: Strategies for Conservation*. Cambridge: Cambridge University Press.

Silber, G. K., A. S. M. Vanderlaan, A. T. Arcevedillo, L. Johnson, C. T. Taggart, M. W. Brown, S. Bettridge, and R. Sagarminaga. 2012. "The Role of the International Maritime Organization in Reducing Vessel Threat to Whales: Process, Options, Action and Effectiveness." *Marine Policy* 36(6):1221–33.

Silber, Gregory K., and Shannon Bettridge. 2012. *An Assessment of the Final Rule to Implement Vessel Speed Restrictions to Reduce the Threat of Vessel Collisions with North Atlantic Right Whales*. National Marine Fisheries Service, National Oceanic and Atmospheric Administration, U.S. Department of Commerce. NOAA Technical Memorandum NMFS-OPR-48. Silver Spring, MD.

Sinclair, Upton. 1927. *Oil*. Boston: Albert & Charles Boni.

Sloan, N. A. 2002. "History and Application of the Wilderness Concept in Marine Conservation." *Conservation Biology* 16(2):294–305.

Snyder, Gary. 1990. *The Practice of the Wild.* San Francisco, CA: North Point Press.

———. 1993. "Mother Earth: Her Whales." In *No Nature.* New York: Pantheon.

———. 1995a. "Coming into the Watershed." *Wild Earth* (special issue) 2:65–70.

———. 1995b. *A Place in Space: Ethics, Aesthetics, and Watersheds: New and Selected Prose.* Washington, DC: Counterpoint.

———. 1996. "Reinhabitation." In his *A Place in Space,* 183–91. Washington, DC: Counterpoint.

Southall, B.L., A.E. Bowles, W.T. Ellison, J.J. Finneran, R.L. Gentry, C.R. Greene Jr., D. Kastak, D.R. Ketten, J.H. Miller, P.E. Nachtigall, W.J. Richardson, J.A. Thomas, and P.L. Tyack. 2007. "Marine Mammal Noise Exposure Criteria: Initial Scientific Recommendations." *Aquatic Mammals* 33(4). doi:10.1578/AM.1533.1574.2007.1411

Southall, Brandon L., David Moretti, Bruce Abraham, John Calambokidis, Stacy L. DeRuiter, and Peter L. Tyack. 2012. "Marine Mammal Behavioral Response Studies in Southern California: Advances in Technology and Experimental Methods." *Marine Technology Society Journal* 46(4):48–59.

Sponsel, Leslie. "Is Indigenous Spiritual Ecology a New Fad?" In *Reflections of Hawai'i Indigenous Traditions and Ecology: The Interbeing of Cosmology and Community,* ed. J. Grim, 159–74. Cambridge, MA: Harvard University Press.

Statistics New Zealand. 2008. "Measuring New Zealand's Progress Using a Sustainable Development Approach: 2008. Topic 2: Biodiversity." www.stats.govt.nz/browse_for_stats/environment/sustainable_development/sustainable-development/biodiversity.aspx.

Steffen, W., K. Richardson, J. Rockstrom, S.E. Cornell, I. Fetzer, E.M. Bennett, R. Biggs, S.R. Carpenter, W. de Vries, C.A. de Wit, C. Folke, D. Gerten, J. Heinke, G.M. Mace, L.M. Persson, V. Ramanathan, B. Reyers, and S. Sorlin. 2015. "Planetary Boundaries: Guiding Human Development on a Changing Planet." *Science* 347(6223). www.sciencemag.org/content/347/6223/1259855.full.

Stein, B.A., L.S. Kutner, and J.S. Adams, eds. 2000. *Precious Heritage: The Status of Biodiversity in the United States.* New York: Oxford University Press.

Stein, Paul. 2001. *Biomes of the Future.* New York: Rosen.

Stiles, Kristin. 1996. "Performance Art." In *Theories and Documents of Contemporary Art: A Sourcebook of Artists' Writing,* ed K. Stiles and P. Selz, 679–94. Berkeley: University of California Press.

Taylor, Rowan. 1997. *The State of New Zealand's Environment 1997.* Wellington: Ministry for the Environment.

Tarnas, Richard. 1992. *The Passion of the Western Mind: Understanding Ideas That Have Shaped Our World View.* New York: Ballantine.

Taussig, Michael. 1993. *Mimesis and Alterity: A Particular History of the Senses.* London: Routledge.

Thayer, Robert L. Jr. 2003. *Life Place: Bioregional Thought and Practice.* Berkeley: University of California Press.

Thoreau, Henry D. 1854. *Walden.* Boston: Ticknor and Fields.

Townsend, Colin R., and Todd A. Crowl. 1991. "Fragmented Population Structure in a Native New Zealand Fish: An Effect of Introduced Brown Trout?" *Oikos* 61:347–54.

Tuan, Yi-Fu. 2001. *Space and Place: The Perspective of Experience.* Minneapolis: University of Minnesota Press.

Turner, Frederick. 1991. *Beauty: The Value of Values.* Charlottesville: University Press of Virginia.

Turnipseed, Mary, Stephen E. Roady, Raphael Sagarin, and Larry B. Crowder. 2009. "The Silver Anniversary of the United States' Exclusive Economic Zone: Twenty-Five Years of Ocean Use and Abuse, and the Possibility of a Blue Water Public Trust Doctrine." *Ecology Law Quarterly* 36:1–70.

Tuwhare, Hone. 1964. "Not by Wind Ravaged." In *No Ordinary Sun.* Auckland: Blackwood and Janet Paul.

———. 1975. "A Tail for Maui's Wife." In *Something Nothing.* Dunedin: Caveman.

United Nations Environmental Programme. 2013. *New Awareness of Opportunities for UNEP to Address Climate Change in the Arctic.* Nairobi.

US Department of Commerce, National Oceanic and Atmospheric Administration, National Marine Sanctuary Program. 2007. *Channel Islands National Marine Sanctuary Final Environmental Impact Statement for the Consideration of Marine Reserves and Marine Conservation Areas.* Silver Spring, MD.

Vanderlaan, Angelia S.M., and Christopher T. Taggart. 2007. "Vessel Collisions with Whales: The Probability of Lethal Injury Based on Vessel Speed." *Marine Mammal Science* 23(1):144–56.

Vince, Joanna, and Marcus Haward. 2009. "New Zealand Oceans Governance: Calming Turbulent Waters?" *Marine Policy* 33(2):412–18.

Volkerling, Kier. 2006. *Kaitiakitanga and Integrated Management.* Unpublished manuscript on file with the author.

Watson, Reg, and Daniel Pauly. 2001. "Systematic Distortions in World Fisheries Catch Trends." *Nature* 414(6863):534–36.

Weaver, Helen. 1976. *Antonin Artaud: Selected Writings.* New York: Farrar, Straus, and Giroux.

Wedde, Ian. 1975. *Pathway to the Sea.* Christchurch, NZ: Hawk Press.

Weible, Christopher M. 2008. "Caught in a Maelstrom: Implementing California Marine Protected Areas." *Coastal Management* 36(4):350–73.

Weible, Christopher M., and Paul A. Sabatier. 2005. "Comparing Policy Networks: Marine Protected Areas in California." *Policy Studies Journal* 33(2):181–201.

Wermund, Edward G. 1985. "Historic Leasing, Exploration and Development Activity in Texas State Waters." In *Proceedings, Fifth Annual Gulf of Mexico Information Transfer Meeting*, 284–89. OCS Study MMS 85–0008. U.S. Dept. of Interior, Minerals Management Service, Gulf of Mexico OCS Region, New Orleans, LA.

Western Water Policy Review Advisors Commission. 1998. *Water in the West: Challenges for the Next Century. Final Report.* Washington, DC.

White, C., B.S. Halpern, and C.V. Kappel. 2012. "Ecosystem Service Tradeoff Analysis Reveals the Value of Marine Spatial Planning for Multiple Ocean Uses." *Proceedings of the National Academy of Sciences* 109(12):4696–4701.

White, Richard. 1995. *The Organic Machine: The Remaking of the Columbia River.* New York: Hill & Wang.

Whittaker, Robert J., and José María Fernandez-Palacios. 2007. *Island Biogeography: Ecology, Evolution, and Conservation.* Oxford: Oxford University Press.

Whyte, William Hollingsworth. 1958. *The Exploding Metropolis.* New York: Doubleday.

Wilcove, David S., David Rothstein, Jason Dubow, Ali Phillips, and Elizabeth Losos. 2000. "Threats to Imperiled Species in the United States." *BioScience* 48:607–15.

Wiley, David N., Michael Thompson, and Richard M. Pace. 2011. "Modeling Speed Restrictions to Mitigate Lethal Collisions between Ships and Whales in the Stellwagen Bank National Marine Sanctuary, USA." *Biological Conservation* 144(9):2377–81.

Woolley, John T., and Michael V. McGinnis. 1999. "The Politics of Watershed Policymaking: Three Cases Compared." *Policy Studies Journal* 27(3):578–98.

———. 2000. "The Conflicting Discourses of Restoration." *Society and Natural Resources* 13:339–57.

———. 2002. "The California Watershed Movement: Science and the Politics of Place." *Natural Resources Journal* 42:133–83.

World Wildlife Fund. 2014. *Living Planet Report 2014: Species and Spaces, People and Places.* Gland, Switzerland: World Wildlife Fund International.

Worm, B., R. Hilborn, J.K. Baum, T.A. Branch, J.S. Collie, C. Costello, M.J. Fogarty, E.A. Fulton, J.A. Hutchings, S. Jennings, O.P. Jensen, H.K. Lotze, P.M. Mace, T.R. Mcclanahan, C. Minto, S.R. Palumbi, A.M. Parma, D. Ricard, A.A. Rosenberg, R. Watson, and D. Zeller. 2009. "Rebuilding Global Fisheries." *Science* 325(5940):578–85.

Worster, Donald. 1979. *Nature's Economy: Garden City.* New York: Anchor Press/Doubleday.

———. 1992. *Rivers of Empire: Water, Aridity, and Growth of the American West.* London: Oxford University Press.

Young, David. 2006. *Woven by Water: Histories from the Whanganui River.* Wellington, NZ: Huia Publishers.

Young, Oran A., Gail Osherenko, Julia Ekstrom, Larry B. Crowder, John C. Ogden, and James A Wilson. 2007. "Solving the Crisis in Ocean Governance: Place-Based Management of Marine Ecosystems." *Environment* 49:8–19.

Zeidberg, Louis D., William M. Hamner, Nikolay P. Nezlin, and Annette Henry. 2006. "The Fishery for California Market Squid (*Loligo opalescens*) (Cephalopoda: Myopsida), from 1981 through 2003." *Fishery Bulletin* 104:46–59.

Index